Cancer Biology: How Science Works

Carsten Carlberg · Eunike Velleuer

Cancer Biology: How Science Works

Carsten Carlberg
Institute of Biomedicine
University of Eastern Finland
Kuopio, Finland

Eunike Velleuer
Department for Pediatric Hemato-Oncology
Children's Hospital
Krefeld, Nordrhein-Westfalen, Germany

ISBN 978-3-030-75698-7 ISBN 978-3-030-75699-4 (eBook)
https://doi.org/10.1007/978-3-030-75699-4

This Springer imprint is published by the registered company Springer Nature Switzerland AG
The registered company address is: Gewerbestrasse 11, 6330 Cham, Switzerland

Preface

This book starts with bad news: one in two of us will face during our lifespan the diagnosis of cancer, i.e., the detection of a malignant tumor. However, the good news are that less than half of all cancer patients die from the disease and that every second **cancer death is preventable**. Cancer is a collection of diseases that can affect basically every organ of our body, all of which have in common uncontrolled cellular growth. Behind every newly diagnosed malignant tumor in adulthood there is an **individual history of probably 20 or more years of tumorigenesis**. Cancer is typically considered as a disease of our genome caused by the accumulation of DNA point mutations as well as translocations, deletions and amplifications of larger genomic regions. However, tumorigenesis also comes along with abnormalities in cellular identity, different responsiveness to internal and external stimuli and major changes in the transcriptome, all of which are based on changes of our epigenome. In fact, most types of cancer carry mutations both in the **genome and epigenome**.

The cells forming our body have the potential to grow in the context of wound healing or for the constant replacement of cells in our blood, skin or intestine. Malignant tumor formation often takes time making cancer in most cases to an **age-related disease** that we seem not to be able to evade. However, tumorigenesis is dependent on multiple environmental influences, including the surveillance of the immune system for cancer cells. Many pro- and anti-cancer effects of the environment we have under control by **lifestyle decisions**, such as retaining from smoking, selecting healthy food and being physically active. Thus, **cancer preventive interventions are the most effective way to fight against cancer**.

We should try to understand ourselves in detail as well as in a global setting. Basic biology explains cellular mechanisms, such as growth, differentiation and cell death, which make life as a whole possible. Every (human) organism represents a complex interplay between **hundreds of different cell types** forming distinctive tissues and organs with specialized tasks. These processes need to be highly orchestrated especially during embryogenesis and later on in maintaining homeostasis of the adult body. Studying the cellular and molecular basis of these mechanisms is one of the most fascinating areas but also a great challenge. Nevertheless, research made the biggest steps in elucidating biological processes via studying malfunctions of normal mechanisms leading to different diseases. Therefore, this textbook wants not only to describe **basic mechanism leading to cancer** but also to provide the readers with a more holistic view and placing these insights in the context of the **personal consequences of everyone's lifestyle decisions**.

The content of the book is linked to the lecture course in "Cancer Biology", which is part of a series together with courses in "Molecular Medicine and Genetics", "Molecular Immunology" and "Nutrigenomics", that is given by one of us (C. Carlberg) in different forms since 2005 at the University of Eastern Finland in Kuopio. Moreover, biological processes explained in this book will be set into a clinical context using the experience of the daily care in oncology. This book also relates to the textbooks "Mechanisms of Gene Regulation: How Science Works" (ISBN 978-3-030-52321-3), "Human Epigenetics: How Science Works" (ISBN 978-3-030-22907-8) and "Nutrigenomics: How Science Works" (ISBN 978-3-030-36948-4), the studying of which may be interesting to readers who like to get more detailed information. The clinical impact of the book is based on personal experience in detecting, treating and preventing cancer by one of us (E. Velleuer).

Chapter 1 of this book will provide a general overview on cancer. Chapters 2–6 focus on the molecular basis of the disease, while Chaps. 7–9 will discuss the cellular aspects of cancer. Finally, Chaps. 10 and 11 will explain the concepts of an efficient therapy against cancer. By combining basic understanding of cellular mechanism of cancer with clinical examples, we hope to make this textbook a personal experience. A glossary in the appendix will explain the major specialist's terms.

We hope that readers will enjoy this rather visual book and get as enthusiastic as the authors about life and its malfunction counterpart reflected in cancer biology.

Kuopio, Finland
Düsseldorf, Germany
March 2021

Carsten Carlberg
Eunike Velleuer

Acknowledgements

The authors would like to dedicate this book to all victims of cancer, whom they know in person. Some managed to survive and enrich the world with their presence. Others are not anymore with us but will never be forgotten.

Contents

Abbreviations

3C	Chromosome conformation capture
3D	3-dimensional
5hmC	5-hydroxymethylcytosine
5hmU	5-hydroxyuracil
5mC	5-methylcytosine
A	Adenine
ABL1	ABL proto-oncogene 1, non-receptor tyrosine kinase
ADAMTS13	ADAM metallopeptidase with thrombospondin type 1 motif 13
AIDS	Acquired immune deficiency syndrome
AKT	Akt murine thymoma viral oncogene homolog
ALK	Anaplastic lymphoma kinase
ALL	Acute lymphoid leukemia
AML	Acute myeloid leukemia
AMPK	AMP-activated protein kinase
AP-1	Activating protein 1
APC	APC regulator of WNT signaling pathway
APAF1	Apoptotic peptidase activating factor 1
APOBEC	Apolipoprotein B mRNA editing catalytic subunit
APR	Acute protein response
ARID	AT-rich interaction domain
AR	Androgen receptor
ASNS	Asparagine synthetase (glutamine-hydrolyzing)
ATAC-seq	Assay for transposase-accessible chromatin using sequencing
ATM	ATM serine/threonine kinase
ATR	ATR serine/threonine kinase
ATRX	ATRX chromatin remodeler
BAX	BCL2 associated X, apoptosis regulator
BCL2	BCL2 apoptosis regulator
BCL2L	BCL2 like
BCR	B cell receptor
BER	Base excision repair
BID	BH3 interacting domain death agonist

bp	Base pair
BRAF	B-Raf proto-oncogene, serine/threonine kinase
BRCA1	BRCA1 DNA repair associated
BRD	Bromodomain containing
BRIP1	BRCA1 interacting protein C-terminal helicase 1
C	Cytosine
C1QBP	Complement C1q binding protein
CAD	Carbamoyl-phosphate synthetase
CAR	Chimeric antigen receptor
CCL	Chemokine (C-C motif) ligand
CCN	Cyclin
CCR	C-C chemokine receptor
CD	Cluster of differentiation
CDC7	Cell division cycle 7
CDH	Cadherin
CDK	Cyclin-dependent kinase
CDKI	Cyclin-dependent kinase inhibitor
CDKN2A	Cyclin-dependent kinase inhibitor 2A
CDR	Complementarity-determining region
CDT1	Chromatin licensing and DNA replication factor 1
CEBPA	CCAAT enhancer binding protein alpha
CHEK	Checkpoint kinase 2
ChIP	Chromatin immunoprecipitation
CIMP	CpG island methylator phenotype
CIN	Cervical intraepithelial neoplasia
CIS	Carcinoma *in situ*
CLEC12A	C-type lectin domain family 12 member A
CLL	Chronic lymphoid leukemia
CML	Chronic myeloid leukemia
CNV	Copy number variation
COVID-19	Coronavirus disease 2019
CREBBP	CREB binding protein, also called KAT3A
CSF	Colony stimulating factor
CSNK2A1	Casein kinase 2 alpha 1
CT	Computed tomography
CTCF	CCCTC binding factor
CTLA4	Cytotoxic T lymphocyte associated protein 4
CTNNB1	Catenin beta 1
CXCL	Chemokine (C-X-C motif) ligand
CYP19A1	Cytochrome P450 family 19 subfamily A member 1
DAMP	Damage-associated molecular pattern
DAXX	Death domain associated protein
DBD	DNA-binding domain
DCIS	Ductal carcinoma *in situ*
DDR	DNA damage response

DLBCL	Diffuse large B cell lymphoma
DNase-seq	DNase I hypersensitivity followed by sequencing
DNMT	DNAMethyltransferase
DOT1L	DOT1 like histone lysine methyltransferase
EBV	Epstein-Barr virus
EGF	Epidermal growth factor
EGFL	EGF like
EGFR	Epidermal growth factor receptor
EHMT2	Euchromatic histone lysine methyltransferase 2
EIF	Eukaryotic translation initiation factor
ELK1	ETS transcription factor ELK1
EMA	European Medicines Agency
EMT	Epithelial-mesenchymal transition
ENCODE	Encyclopedia of DNA elements
EP300	E1A binding protein p300, also called KAT3B
ERBB2	Erb-B2 receptor tyrosine kinase 2, also called HER2
ERCC3	ERCC excision repair 3, TFIIH core complex helicase subunit
eRNA	Enhancer RNA
ES	Embryonic stem
ESR1	Estrogen receptor
EV	Extracellular vesicle
EZH	Enhancer of zeste homolog
FACS	Fluorescence-activated cell sorting
FAD	Flavin adenine dinucleotide
FAP	Familial adenomatous polyposis
FASN	Fatty acid synthase
FBXO32	F-box protein 32
FDA	US Food & Drug Administration
FGF	Fibroblast growth factor
FLT	Fms related receptor tyrosine kinase
FOS	Fos proto-oncogene, AP-1 transcription factor subunit
FOXO1	Forkhead box O1
G	Guanine
GART	Phosphoribosylglycinamide formyltransferase
GDF	Growth differentiation factor
GDP	Guanosine diphosphate
GLS	Glutaminase
GR	Glucocorticoid receptor
GRB2	Growth factor receptor bound protein 2
GSH	Glutathione
GTF2H4	General transcription factor IIH subunit 4
GTP	Guanosine triphosphate
GWAS	Genome-wide association study
HAT	Histone acetyltransferase
HBV	Hepatitis B virus

HDAC	H istone deacetylase
HGF	Hepatocyte growth factor
Hi-C	High-throughput chromosome capture
HIF1A	Hypoxia inducible factor 1 subunit alpha
HIV	Human immunodeficiency virus
HK2	Hexokinase 2
HLA	Human leukocyte antigen
HP1	Heterochromatin protein 1, official name CBX5
HPV	Human papilloma virus
HR	Homologous recombination
HSP	Heat shock protein
ICAM	Intercellular adhesion molecule
ICGC	International Cancer Genome Consortium
IDH	Isocitrate dehydrogenase
IDO1	Indoleamine 2,3-dioxygenase 1
Ig	Immunoglobin
IGF	Insulin-like growth factor
IGH	Immunoglobulin heavy locus
IL	Interleukin
IL3RA	Interleukin 3 receptor subunit alpha
IL6R	IL6 receptor
indel	Short insertion or deletion
INFγ	Interferon γ
iPS	Induced pluripotent stem
ITAM	Immunoreceptor tyrosine-based activation motif
JAK	Janus kinase
JUN	Jun proto-oncogene, AP-1 transcription factor subunit
kb	Kilo base pairs (1000 bp)
KDM	Lysine demethylase
KDR	Kinase insert domain receptor, also called VEGFR2)
KIT	KIT proto-oncogene, receptor tyrosine kinase
KLF4	Krüppel-like factor 4
KMT	Lysine methyltransferase
LAD	Lamin-associated domain
LDHA	Lactate dehydrogenase A
LIN28A	Lin-28 homolog A
LINE	Long interspersed element
LOCK	Large organized chromatin K9-modification
LPS	Lipopolysaccharide
LSD1	Lysine specific demethylase 1, also called KDM1A
LTR	Long terminal repeat
MAF	Minor allele frequency
MAP2K	Mitogen-activated protein kinase kinase
MAPK	Mitogen-activated protein kinase
MAX	MYC associated factor X

Mb	Mega base pairs (1,000,000 bp)
MBD	Methyl-DNA binding domain
MCM	Minichromosome maintenance complex component
MDM2	MDM2 proto-oncogene, E3 ubiquitin protein ligase
MDSC	Myeloid-derived suppressor cell
MECP2	Methyl-CpG binding protein 2
MET	Mesenchymal-epithelial transition
MGMT	O-6-methylguanine-DNA methyltransferase
MHC	Major histocompatibility complex
miRNA	Micro RNA
MIS-C	Multisystem inflammatory syndrome in children
MLH1	MutL homolog 1
MMP	Matrix metalloproteinase
MMR	Mismatch repair
MNT	MAX network transcriptional repressor
MRI	Magnetic resonance imaging
mRNA	Messenger RNA
MS	Myeloid sarcoma
MSI	Microsatellite instability
MSS	Microsatcllite stable
mTOR	Mammalian target of rapamycin
MYC	MYC proto-oncogene, BHLH transcription factor
NAD	Nicotinamide adenine dinucleotide
NADPH	Nicotinamide adenine dinucleotide phosphate
NANOG	Nanog homeobox
ncRNA	Non-coding RNA
NELFE	Negative elongation factor complex member E
NER	Nucleotide excision repair
NF-κB	Nuclear factor κB
NHEJ	Non-homologous end-joining
NK	Natural killer
NLS	Nuclear localization sequence
NO	Nitric oxide
NPM1	Nucleophosmin 1
NSD	Nuclear receptor binding SET domain protein
nt	Nucleotides
OCT4	Octamer-binding transcription factor 4
PALB2	Partner and localizer of BRCA2
PAMP	Pathogen-associated molecular pattern
PCAWG	PanCancer Analysis of Whole Genomes
PDCD1	programmed cell death 1, also called PD1
PDGF	Platelet-derived growth factor
PDGFRA	Platelet-derived growth factor receptor α
PET	Positron emission tomography
PGE2	Prostaglandin E2

PI3K	Phosphoinositide 3-kinase
PICS	PTEN loss-induced cellular senescence
PIK3CA	Phosphatidylinositol-4,5-bisphosphate 3-kinase catalytic subunit alpha
PIN	Prostatic intraepithelial neoplasia
PIP3	Phosphatidylinositol-3,4,5-triphosphate
PMAIP1	Phorbol-12-myristate-13-acetate-induced protein 1
PML	PML nuclear body scaffold
Pol II	RNA polymerase II
PPARGC1B	PPARG coactivator 1 beta
PPAT	Phosphoribosyl pyrophosphate amidotransferase
PRC	Polycomb repressive complex
PRKDC	Protein kinase, DNA-activated, catalytic subunit
PRMT5	Protein arginine methyltransferase 5
PTEN	Phosphatase and tensin homolog
PTHrP	Parathyroid hormone-related protein
RAF1	Raf-1 proto-oncogene, serine/threonine kinase
RARA	Retinoic acid receptor alpha
RAS	Rat sarcoma
RB1	RB transcriptional corepressor 1
ROS	Reactive oxygen species
RRM2	Ribonucleotide reductase regulatory subunit M2
rRNA	Ribosomal RNA
RSV	Rous sarcoma virus
RTK	Receptor tyrosine kinase
SAM	S-adenosyl-L-methionine
SARS-CoV 2	severe acute respiratory syndrome coronavirus 2
SCD	Stearoyl-CoA desaturase
scFv	Single-chain variable fragment
SERPINE1	Serpin peptidase inhibitor, clade E
SETD2	SET domain containing 2
SETDB1	SET domain bifurcated histone lysine methyltransferase 1
SHC1	SHC adaptor protein 1
SHMT	Serine hydroxymethyltransferase
SINE	Short interspersed element
SIRT	Sirtuin
SLAMF7	SLAM family member 7
SLC	Solute carrier family
SMARC	SWI/SNF-related matrix-associated actin-dependent regulators of chromatin
SNAI	Snail family transcriptional repressor
snoRNA	Small nucleolar RNA
SNP	Single nucleotide polymorphism
SNV	Single nucleotide variant
SOS1	SOS Ras/Rac guanine nucleotide exchange factor 1

SOX2	SRY-box 2
SP1	Specificity protein 1
SRC	SRC proto-oncogene, non-receptor tyrosine kinase
SRF	Serum response factor
STAB1	Stabilin
STAT	Signal transducer and activator of transcription
SUV39H1	Suppressor of variegation 3-9 homolog 1, also called KMT1A
SV40	Simian virus 40
SWI/SNF	Switching/sucrose non-fermenting
T	Thymine
TAM	Tumor-associated macrophage
TCGA	The Cancer Genome Atlas
TCR	T cell receptor
TDG	Thymine-DNA glycosylase
TERT	Telomerase reverse transcriptase
TET	Ten-eleven translocation
TFAM	Transcription factor A, mitochondrial
TGFβ	Transforming growth factor β
TGFBR	TGFβ receptor
T_H	T helper
THBS1	Thrombospondin 1
TIL	Tumor infiltrating lymphocyte
Tis	Carcinoma *in situ*
TLR4	Toll-like receptor 4
TNF	Tumor necrosis factor
TNFRSF	TNF receptor superfamily member
TP53	Tumor protein p53
T_{reg}	T regulatory
TRIM	Tripartite motif containing
tRNA	Transfer RNA
TSS	Transcription start site
TWIST1	Twist family BHLH transcription factor 1
U	Uracil
UBTF	Upstream binding transcription factor
UGDH	UDP-glucose 6-dehydrogenase
UICC	Union for International Cancer Control
VEGF	Vascular endothelial growth factor
VHL	Von Hippel-Lindau tumor suppressor
WAT	White adipose tissue
WHO	World Health Organization
WNT	Wingless-type MMTV integration site family member
ZEB	Zinc finger E-box binding homeobox

Chapter 1
Introduction to Cancer

Abstract Cancer is the second leading cause of death that globally kills nearly 10 million people every year. One in two of us will be diagnosed with cancer at some point of his/her life. Malignant tumors can arise from different tissues and organs, thus there are many different types of cancer. In all of them the normal control of cell division and differentiation is lost, so that an individual cell multiplies inappropriately forming a primary malignant tumor. The cancer cells may eventually spread through the body and form potentially deadly metastases. The most promising strategy for reducing both the number of cancer cases as well as their mortality are preventive interventions like eliminating exposure to carcinogens, such as tobacco smoke.

Keywords Non-communicable diseases · Mortality rates · Cancer driver genes · Tumorigenesis · Cancer prevention

1.1 The Global Burden of Cancer

Non-communicable diseases contribute to more than 73% of deaths worldwide and even to a higher percentage in industrialized countries. In contrast, infectious (communicable), maternal, neonatal and nutritional diseases account only for less than 19% of worldwide deaths and injuries for 8%. In 2017 worldwide 17.8 million persons were dying from cardiovascular diseases, while **neoplasms killed 9.6 million humans representing 17.1% of worldwide deaths** (Fig. 1.1, top left). For comparison, in an industrialized country, such as Finland, the rate of cancer death was even 24.4% (Fig. 1.1, bottom left).

In the past, cancer was seldomly the cause of death, since the average human life expectancy was far shorter and people died from other causes than cancer. However, medical care significantly changed during the last 100 years leading to drastic improvements in life expectancy. Unfortunately, this caused an increase in the overall burden of cancer and accordingly to a far higher number of cancer cases. In

C. Carlberg and E. Velleuer, *Cancer Biology: How Science Works*,
https://doi.org/10.1007/978-3-030-75699-4_1

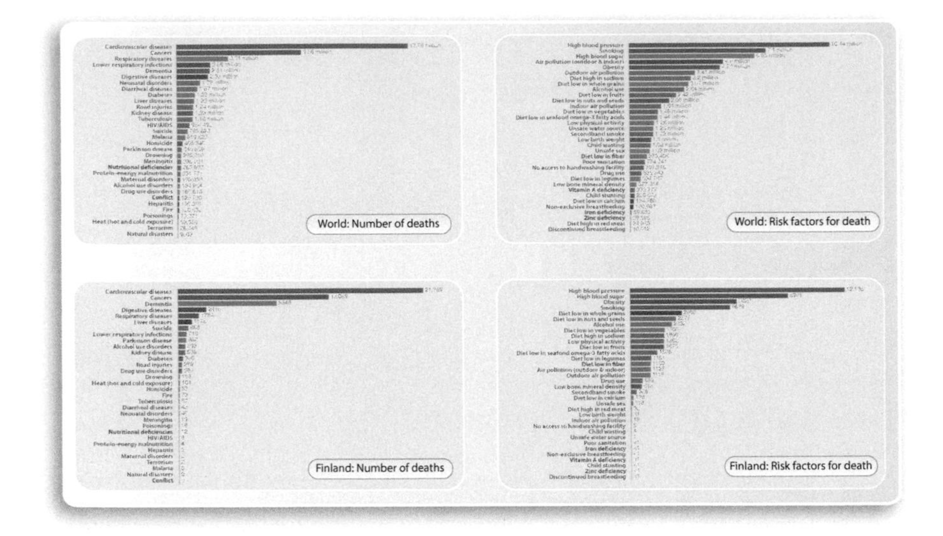

Fig. 1.1 Leading causes of death. The number of deaths are listed by cause (**left**) and by risk factor (**right**). The worldwide data (**top**) and are compared based on data from Finland (**bottom**) are based on https://ourworldindata.org/causes-of-death and describe the year 2017. In that year worldwide 56 million people died and in Finland 53,700

2017, the leading type of lethal cancer was lung and broncheal cancer with 1.9 million yearly victims followed by colorectum cancer (0.9 million deaths). Importantly, both groups of cancer are to 70–90% preventable (Sect. 1.4). Accordingly, smoking, obesity and unhealthy diet belong to the major risk factors for death, both worldwide and in Finland (Fig. 1.1, right). Importantly, **tobacco smoking is the single most important cancer-related lifestyle factor**, stopping of which would have the largest benefit, not only on the rate of cancer, but it also would significantly reduce the risk of cardiovascular, lung and kidney diseases.

The progress of the fight against cancer is more efficiently measured by mortality rate than by incidence or survival rate. Comparing the data of the years 2017 and 2007, the mortality rates of stomach cancer, Hodgkin lymphoma and esophageal cancer decreased by 17.1, 16.8 and 14.5%, respectively. Historically, **during the 20th century cancer death rates significantly increased, primarily due to the tobacco epidemic, and reached its maximum in 1991** with 215 deaths per 100,000 population but is now 29% lower (152 deaths per 100,000 in 2017). This decrease is primarily due to declines in smoking and improvements in early detection and therapy. During the past 10 years the mortality rate of cancer declined in average by 1.5% per year, which mainly reflects the decrease in deaths from the four major types of cancer (lung, breast, prostate and colorectum). Since 1990 lung cancer mortality declined by 51% for males and since 2002 by 26% for females, since 1989 breast cancer mortality is reduced by 40%, since 1993 prostate cancer mortality decreased

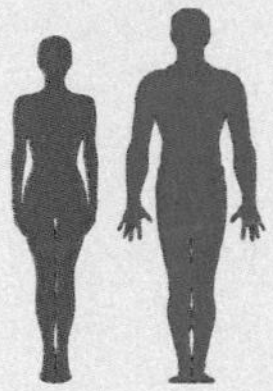

New cases

Females		
Breast	276,480	30%
Lung & bronchus	112,520	12%
Colon & rectum	69,650	8%
Uterine corpus	65,620	7%
Thyroid	40,170	4%
Melanoma of the skin	40,160	4%
Non-Hodgkin lymphoma	34,860	4%
Kidney & renal pelvis	28,230	3%
Pancreas	27,200	3%
Leukemia	25,060	3%
All sites	**912,930**	**100%**

Males		
Prostate	191,930	21%
Lung & bronchus	116,300	13%
Colon & rectum	78,300	9%
Urinary bladder	62,190	7%
Melanoma of the skin	60,190	7%
Kidney & renal pelvis	45,520	5%
Non-Hodgkin lymphoma	42,380	5%
Oral cavity & pharynx	38,380	4%
Leukemia	35,470	4%
Pancreas	30,400	3%
All sites	**893,660**	**100%**

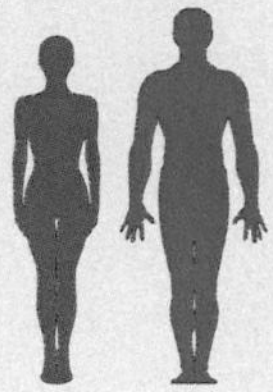

Deaths

Females		
Lung & bronchus	63,220	22%
Breast	42,170	15%
Colon & rectum	24,570	9%
Pancreas	22,410	8%
Ovary	13,940	5%
Uterine corpus	12,590	4%
Liver & intrahepatic bile duct	10,140	4%
Leukemia	9680	3%
Non-Hodgkin lymphoma	8480	3%
Brain & other nervous system	7830	3%
All sites	**285,360**	**100%**

Males		
Lung & bronchus	72,500	23%
Prostate	33,330	10%
Colon & rectum	28,630	9%
Pancreas	24,640	8%
Liver & intrahepatic bile duct	20,020	6%
Leukemia	13,420	4%
Esophagus	13,100	4%
Urinary bladder	13,050	4%
Non-Hodgkin lymphoma	11,460	4%
Brain & other nervous system	10,190	3%
All sites	**321,160**	**100%**

Fig. 1.2 The leading cancer types by diagnosis, deaths and gender. The 10 leading types of cancer based on new cases (**top**) and deaths (**bottom**) for females (**left**) and males (**right**) in the US. Most of the cancers derive out of the epithelial compartment. Notably, the two most common types of cancers (prostate and lung for men and breast and lung in women) are responsible for one third of all cancer deaths. The majority of lung cancers could be avoided by reducing the exposure to tobacco smoke. Data are for the US, 2020, based on Siegel et al. (2020) CA Cancer J Clin *70*, 7–30

by 52% and since 1980 colorectal cancer mortality was 53 and 57% lower for males and females, respectively (Fig. 1.2).

The term "tumor" is a Latin word meaning swelling, i.e., a tumor per se is a growing cell mass, whereas a malignant tumor is also called cancer. The incidence of malignant tumors is significantly varying between human organs. While cancer in colorectum and breast is common, there are hardly any cases in the heart. Moreover, the uterus frequently shows benign hyperplasia, such as fibroids, that rarely become malignant, while the adrenal gland also seldomly develops primary malignant tumors but the organ is a very common target of lung cancer metastases (Sect. 9.3). An organ-specific cancer prevalence is primarily explained by the relative roles of intrinsic and extrinsic risk factors. Accordingly, primary lung cancer is rare in non-smokers but common in smokers (Sect. 4.1). In the same way, primary liver cancer is frequent after tumor virus infections (Box 2.1), alcoholism or chronic iron overload but rare in normal livers. Moreover, two-thirds of the variation in cancer risks may be explained by the total number of stem cell divisions per tissue type, so that **heavily growing tissues, such as the colon, have a higher risk to accumulate mutations based on uncorrected replication errors** (Sect. 4.4).

Another aspect is provided by an evolutionary view, where human fitness is more affected by perturbations in the heart and brain than in the liver. Moreover, paired organs, such as lung and kidney, may stay active even when one of the pair is less functional due to a malignant tumor. Accordingly, the different vulnerabilities to malignant tumors reflects how essential each organ is for survival and reproduction. Therefore, in relatively small organs, such as heart, brain and uterus, it is more important to suppress even small tumors, while this is less important in large or paired organs. For this reason, heart and brain may have developed a very low stem cell growth rate compared with, e.g., the colon. In contrast, in non-essential organs there could be not enough selective pressure to eliminate non-invasive malignant tumors. This may explain why the majority of elderly men carry slowly growing cancers in their prostate. Since mechanisms that are able to suppress the growth of small malignant tumors would also inhibit wound healing (Sect. 8.1), immune function (Chap. 10) and developmental processes that need active cell proliferation, they may not have established. Thus, **cancer is an intrinsic risk of life**.

Approximately, **one third of cancer deaths are due to lifestyle choices, such a high body mass index, low fruit and vegetable intake, lack of physical activity, tobacco and/or alcohol use**. However, only a few definite relationships between specific nutrient-related factors and cancer are established. For example, there is convincing evidence that overweight and obese individuals have increased risk of cancer of esophagus, colorectum, breast (in post-menopausal women), pancreas, liver and kidney, while individuals who consume a high amount of alcohol are prone to cancers of the oral cavity, pharynx, larynx, esophagus, liver, colorectum and breast. Individuals who ingest high amounts of processed meat have increased risk of colorectum cancer (Table 1.1). Additional dietary factors that increase cancer risk include salt-preserved foods (stomach) and food with high glycemic index (endometrium). Moreover, aflatoxins contribute to the development of liver cancer. In contrast, protective factors are coffee (liver and endometrium cancers) and foods high in dietary fiber, such as whole grain products, fruits and vegetables (colorectal cancer). Importantly, **there is convincing evidence that physical activity decreases the risk of colorectal and breast cancer** (endometrium and breast cancer in post-menopausal women).

1.2 Categorization and Diagnosis of Tumors

Our body is composed of some 30 trillion ($3x10^{13}$) cells forming more than 400 different tissues and cell types. These different cell types are the result of embryonic development starting with fertilization, where the two types of haploid gametes, oocyte and sperm, fuse and form the diploid zygote. A series of cleavage divisions provide the blastocyst stage (50–150 cells), where a first differentiation occurs: the outer cells (trophoblasts) form the placenta and other extraembryonic tissues, while the inner cells begin to differentiate into two layers, the epiblast and the hypoblast. The epiblast gives rise to some extraembryonic tissues as well as to all

Table 1.1 Overview of lifestyle factors and risk of developing cancer

Evidence	Decreased risk	Increased risk
Convincing[a]	Physical activity (colon)	• Overweight and obesity (esophagus, colorectum, pancreas, liver, breast in post-menopausal women, endometrium, kidney) • Alcohol (oral cavity, pharynx, larynx, esophagus, liver, colorectum, breast post-menopausal) • Processed meat (colorectum) • Aflatoxin (liver)
Probable[a]	Fruits and vegetables (oral cavity, esophagus, stomach, colorectum[b]) Physical activity (breast in post-menopausal women, endometrium) Whole grain, fiber, dairy products (colorectum) Coffee (liver, endometrium) Alcohol (kidney)	• Red meat (colorectum) • Salt-preserved foods and salt (stomach) • Alcohol (breast in pre-menopausal women) • Chinese-style salted fish (nasopharynx) • Glycemic load (endometrium)
Possible/ insufficient	Fiber Soya Fish ω-3 Fatty acids Carotenoids Vitamins B_2, B_6, folate, B_{12}, C, D, E Calcium, zinc and selenium Non-nutrient plant constituents (e.g., allium compounds, flavonoids, isoflavones, lignans)	• Animal fats • Heterocyclic amines • Polycyclic aromatic hydrocarbons • Nitrosamines

[a]The "convincing" and "probable" categories in this report is taken from the WCRF network report Recommendations and public health and policy implications 2018
[b]For colorectal cancer, a protective effect of fruit and vegetable intake has been suggested by many case-control studies but this has not been supported by results of several large prospective studies, suggesting that if a benefit does exist it is likely to be modest

cells of the later stage embryo and fetus, while the hypoblast is exclusively devoted to making extraembryonic tissues including the placenta and the yolk sac. Some of the embryonic epiblast cells form primordial germ cells, which are the founders of the germline. During the gastrulation phase the other cells of the embryonic epiblast turn into the **three germ layers ectoderm, mesoderm and endoderm** that are the precursors of all somatic tissues. The cells of these germ layers are only multi-potent, i.e., they cannot differentiate into every other tissue. For example, in a series of sequential differentiation steps, ectoderm cells can form epidermis, neural tissue or neural crest, but not kidney (mesoderm-derived) or liver cells (endoderm-derived).

Each cell of the embryo (and the adult) contains an identical diploid genome (Box 1.1), the packaging of which into chromatin serves as a specific filter of information and determines which genes are expressed and which not (Sect. 6.1). The differentiation program of embryogenesis is a perfect system for observing the coordination of cell lineage commitment and cell identity specification. In no other phase of life there is more rapid growth, i.e., **in general, a tumor is not growing as fast as an embryo**. However, in embryogenesis there is coordination between the increase in cellular mass and the phenotypic diversification of the expanding cell populations, while this does not happen during tumorigenesis. Thus, **the principles of embryogenesis demonstrate the proliferation and differentiation potential of our cells for a perfect purpose, while tumorigenesis is misusing these processes and mechanisms**.

Box 1.1: The human genome. The human genome is the complete sequence of the anatomically modern human (*Homo sapiens*) and was studied by the *Human Genome Project* (www.genome.gov/human-genome-project). The human genome sequence that we find in the internet (www.ncbi.nlm.nih.gov/genome/guide/human) as reference genome represents the assembly of the genomes of a few young healthy donors. With the exception of germ cells, i.e., female oocytes and male sperm, platelets and erythrocytes, each human cell contains a diploid genome formed by 2 × 3235 Mb, that is distributed on 2 × 22 autosomal chromosomes and two X chromosomes for females and a XY chromosome set for males. In addition, every mitochondrion contains 16.6 kb mitochondrial DNA. The haploid human genome encodes for some 20,000 protein-coding genes and about the same number of non-coding RNA (ncRNA) genes. The protein-coding sequence covers approximately 1.5% of the human genome, i.e., **the vast majority of the genome is non-coding and seems to have primarily regulatory function**. Almost 50% of the sequence of the human genome is formed by repetitive DNA, which is sorted into the following categories (by order of frequency):

Long interspersed elements (LINEs, 500–8000 bp) 21%
Short interspersed elements (SINEs, 100–300 bp) 11%
Retrotransposons, such as long terminal repeats (LTRs, 200–5000 bp) 8%

DNA transposons (200–2000 bp) 3%
Microsatellite, minisatellite or major satellite (2–100 bp) 3%

LINEs and SINEs are identical or nearly identical DNA sequences that are separated by large numbers of nucleotides, i.e., the repeats are spread throughout the whole human genome. LTRs are characterized by sequences that are found at each end of retrotransposons. DNA transposons are full-length autonomous elements that encode for a transposase, i.e., an enzyme that transposes DNA from one to another position in the genome. Microsatellites are often associated with centromeric or peri-centromeric regions and are formed by tandem repeats of 2–10 bp in length. Minisatellites and major satellites are longer, with a length of 10–60 bp or up to 100 bp, respectively.

The classification of cancer aims to determine the origin of a malignant tumor, i.e., from which cell type or tissue it arose. This precise categorization is essential for efficient communication between oncologist, radiologist, radiation oncologist and surgeons. This uniform system enables the design of treatment protocols, cancer stratifications as well as an international exchange among treating physicians. Furthermore, the classification of cancers is the basis for the correct guidance of the patient through the treatment process as well as for estimating the prognosis. Cancer classification based on the body site where the cancer first developed (anatomical side) is rather inaccurate, since most organs are a composition of different tissues (histological type). For example, there are seven different types of renal cancer in addition to some rare subtypes, all bearing different prognosis and treatment options. Thus, **the classification of cancers is not a semantical task but reflects the understanding of the biological behavior of the malignant tumor leading to the correct therapy.**

The World Health Organization (WHO) and the Union for International Cancer Control (UICC) frequently update the internationally accepted classification of malignant tumors based on histotypes (Fig. 1.3). Details on clinical cancer staging are described in Box 4.2. For solid malignant tumors, there are three major cancer types:

- **carcinomas** (80–90% of all adult cancers) arise out of epithelial cells (e.g., hepatocellular carcinoma or squamous cell carcinoma)
- **blastomas** (1% of all) evolve out of immature precursor cells or embryonic tissues (e.g., neuroblastoma or medulloblastoma) and are most often found in children
- **sarcomas** (1% of all) emerge within connective tissue (e.g., osteosarcoma, Ewing sarcoma) and are often found in young people (beginning of puberty going along with the growth spurt).

The so-called liquid tumors (5% of all cancers in adults) origin in the hematopoietic system. Leukemias are classified by their growth rate into acute or chronic and to their lineage origin in lymphatic or myeloid. Lymphomas also arise out of the hematopoietic system but in contrast to leukemias they infiltrate not more than 20–25% of the bone marrow, i.e., they present with a local swelling of a lymph node or an infiltration in the liver and/or spleen.

Cancer genomics (Chap. 5) and epigenomics (Chap. 6) have revolutionized the classical categorization system, since they describe the molecular characteristics of a cancer. For example, pediatric medulloblastoma (one of the most aggressive form of brain cancers), is nowadays classified based to the defected cancer pathway, such as hedgehog or WNT (wingless-type MMTV integration site family member, Sect. 2.2).

1.3 Crucial Transitions in Cancer

The differentiation of embryonic stem (ES) cells into specialized cell types happens during embryogenesis within the first weeks of the life of a fetus (Sect. 1.2). Moreover,

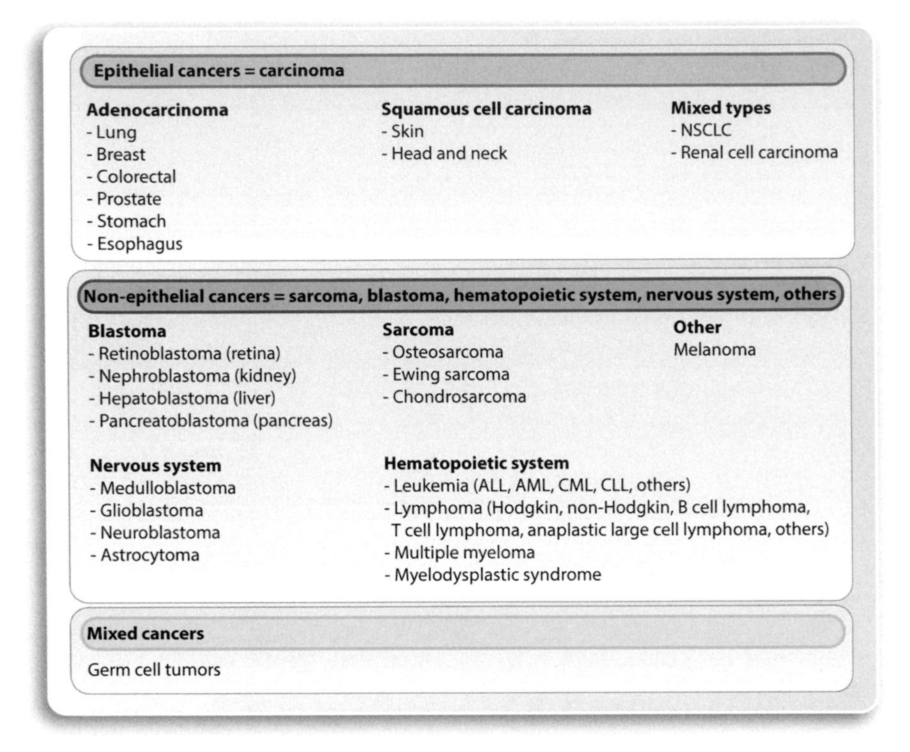

Fig. 1.3 Classification scheme of cancers. The overall histological classification of tumors are in epithelial (**top**), non-epithelial (**center**) and mixed types (**bottom**) tumors. For most malignant tumors a benign counterpart exists (not shown). 80% of all adult cancers are carcinomas and the five leading types of cancer-based deaths are adenocarcinomas (Fig. 1.2). Non-epithelial cancers can be subdivided in at least five different types. 80% of all childhood cancers are non-epithelial cancers (Box 6.3). ALL = acute lymphoid leukemia, AML = acute myeloid leukemia, CLL = chronic lymphoid leukemia, CML = chronic myeloid leukemia, NSCLC = non-small-cell lung cancer

in every moment of an adult's life, stem cells in the bone marrow, the skin, the intestine and other tissues are growing and differentiating into specialized cells that replace cell loss, such as of our immune system or of our body's outer and inner surface. The **epigenetic landscape** (Fig. 1.4, **A**) is a very illustrative model for understanding the underlying molecular mechanisms of cell fate decisions during development. Cellular differentiation may be compared to a system of valleys of a mountain range, where a cell (often represented by a ball), e.g., an ES cell, begins at the top and follows existing paths driven by gravitational force. The latter analogy should express that the path of differentiation has a clear direction. This directs the cell into one of several possible fates represented as valleys that get narrower in the trajectory towards terminally differentiated cell types. Along the downhill path, cell fate decisions need to be taken at bifurcation points. These decisions often depend on the expression of lineage-determining transcription factors. Once a cell has taken a decision, it is restricted in its subsequent decisions by the route it has taken, i.e., **under**

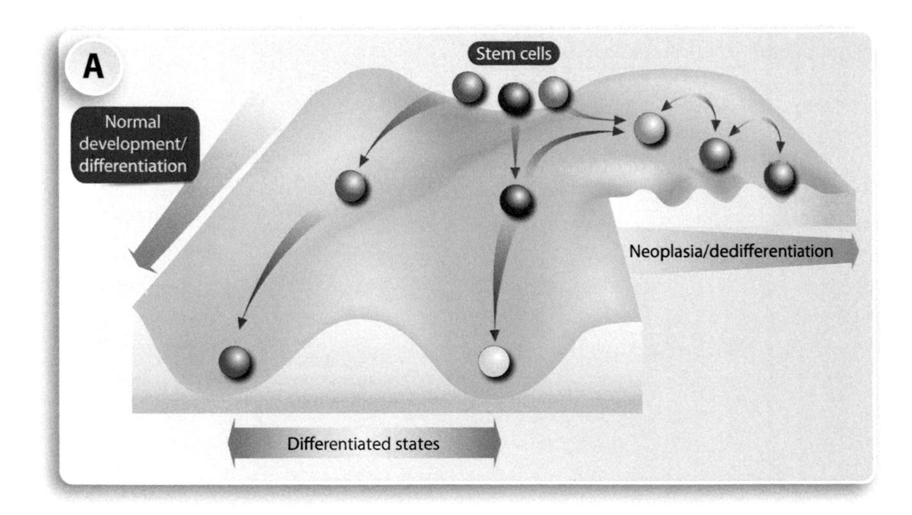

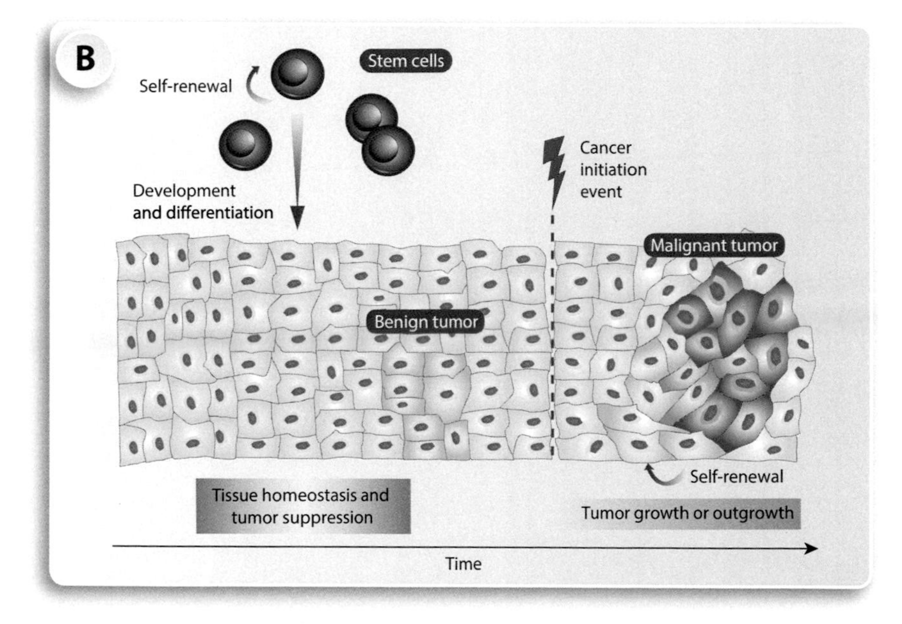

Fig. 1.4 Cellular differentiation and tumorigenesis. Waddington's epigenetic landscape model can be used for the illustration of phenotypic plasticity of cells during normal development as well as of cellular reprogramming in tumorigenesis (**A**). Tumorigenesis is a multi-step process, where first a cancer initiating event transforms a cell into a tumor cell that obtains growth advantages over its neighboring cells (**B**). Over time further "hits" provide the cells with additional hallmarks of cancer that finally lead to outgrowth and metastasis

natural conditions this is an irreversible forward-moving process resulting in highly specialized, terminally differentiated cell types that do not proliferate.

Cancer is characterized by uncontrolled cellular proliferation caused by an initiating event, such as a genetic mutation, an infection with a tumor virus (Sect. 2.1) or the aberrant reactivation of a pluripotency transcription factor (Sect. 6.3). The mutations are called "drivers" when they are able to initiate and progress the process of tumorigenesis (Sect. 4.4). Driver genes are called **oncogenes** (Chap. 2), when they are activated by point mutations, translocations or amplifications, while **tumor suppressor genes** (Chap. 3) are genes that are inactivated by point mutations, deletions or epigenetic silencing. Mutated driver genes affect the homeostasis of the concerned cell and provide it with selective growth advantages over neighboring cells, i.e., they transform the cell. In the epigenetic landscape model, the cell then reaches a state of higher entropy (moving uphill), in which it again proliferates and self-renews, i.e., it is dedifferentiated compared with its normal counterparts (Fig. 1.4, **A** right). In a simplified illustration of tumorigenesis the transformed cell overgrowths its neighboring cells and disturbs tissue homeostasis and architecture (Fig. 1.4, **B** right). Over time the transformed cell or its successors accumulate further events that provide it with additional properties, such as immortalization (Chap. 7), induction of angiogenesis and other changes in the tumor microenvironment (Chap. 8) and finally the potential for evasion and metastasis (Chap. 9). These events are not only genetic mutations in cancer genes (Chap. 5) but also transcriptional and epigenetic changes (Chap. 6).

On the cellular level, tumorigenesis represents a series of transitions from a pre-malignant to a malignant state, from a local malignant tumor to a metastatic disease and from a drug-responsive to a drug-resistant cancer. These transitions involve interactions between a number of pre-malignant, malignant and non-malignant cells of the microenvironment, such as stroma and immune cells (Chap. 8). However, in contrast to embryogenesis, **the initial steps of tumorigenesis cannot easily be observed in the natural in vivo environment**, since tumors have to accumulate billions of cells before they can be detected (Sect. 4.1).

1.4 Causes of Cancer

The widely accepted model of tumorigenesis suggests that cancer develops from a gradual accumulation of driver mutations stepwise increasing the proliferation rate of cancer cells (Fig. 1.5). Tumorigenesis initiates from a normal cell transforming into a tumor cell forming in this way the core of a small benign tumor. Cellular transformation (Sect. 2.1) is triggered either by environmental factors, such as exposure to carcinogens, or by random genetic mutations, which often result from DNA replication errors. The latter is the reason, why rapidly growing tissues, such as colon, have a higher risk of developing cancer (Sect. 4.4). The cancer risk is further enhanced for individuals who carry inherited germline mutations in driver genes (Sect. 5.1). For example, mutation or deletion of the tumor suppressor genes *BRCA1*

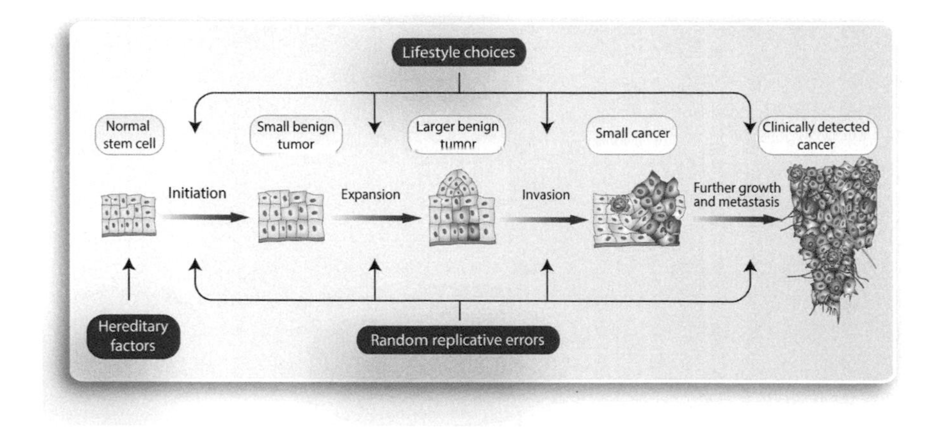

Fig. 1.5 Hereditary, environmental and genetic perspectives on tumorigenesis. In adults, solid malignant tumors develop due to a stepwise accumulation of cellular properties, called hallmarks of cancer, and is referred to as tumorigenesis. Hereditary factors, such as germline mutations, may accelerate this process. Moreover, environmental factors, such as lifestyle choices on smoking, physical activity and diet, modulate the tumorigenesis process, which in its core is based on random genetic mutations that often relate to replication errors

(BRCA1 DNA repair associated) and *BRCA2* significantly increase the risk for gynecological cancers, those of *APC* (APC regulator of WNT signaling pathway) lead to familial adenomatous polyposis (FAP) often causing colorectal cancer and of *RB1* (RB transcriptional corepressor 1) retinoblastoma.

The mutation rate in tissues that are in direct contact with the environment, such as skin, lung or colon, is largely modulated by mutagens like the exposure to UV radiation or by carcinogens contained in tobacco smoke and food. Interestingly, some cancer genes encode for enzymes regulating the metabolism of nutrients and the mutation of these genes can lead to the production of onco-metabolites affecting chromatin modifying enzymes (Sect. 6.1). Accordingly, dietary molecules affecting the epigenetic signaling can both increase or reduce the risk of cancer. Migrant studies showed that moving from a region with low risk for a given cancer to one with a high risk leads within one generation to the same cancer pattern of the host country. Studies with monozygotic twins support these findings. In addition, as already discussed in Sect. 1.1, low physical exercise, low fruit and vegetable intake, high body mass index and alcohol use have an impact on cancer onset. Thus, **environmental exposure and lifestyle choices have a significant contribution to cancer onset**.

Organs that are often exposed to microbes, such the liver by hepatitis B virus (HBV), the stomach by *Helicobacter pylori* bacteria or the cervix by human papillomavirus (HPV), have a higher risk of developing malignant tumors. Moreover, an important step in cancer development is the interaction of pre-cancerous cells with the immune system (Chap. 10). Most common is **chronic inflammation**, which is associated with immune cell invasion into the malignant tumor, the secretion of pro-inflammatory cytokines and changes in tissue organization (Sect. 8.4). The latter involves **angiogenesis**, i.e., a process in which new blood vessels are formed. Additionally, diseases associated with immunosuppression (e.g., HIV (human immunodeficiency virus) infections) also display a higher cancer risk compared to the general population (Sect. 10.1).

Taken together, the tumorigenesis process creates a scenario, in which microbe infection in combination with further external factors, such as high-caloric diet, establish a chronic pro-inflammatory milieu that allows genetic and epigenetic perturbation to take over the control on proliferation and start a malignant tumor. Thus, **the relative cancer risk of a given tissue is based on hereditary, environmental and acquired genetic factors**. The relative contribution of the three drivers of tumorigenesis differ for every person and every case of cancer. For example, based on data from 32 common cancers in the United Kingdom, 29% of driver gene mutations are based on environment factors, 5% on inherited germline mutations and 66% on random replication errors (Sect. 4.4).

1.5 Cancer Prevention

Since cancer is the second most deadly disease worldwide (Sect. 1.1), its prevention should be a key issue of health politics. **Primary prevention**, i.e., avoiding to evolve any malignant tumor in the body, is by far the best strategy. Cancer types that largely occur due to environmental exposure and lifestyle choices, such as lung and colorectal cancer (Sect. 1.4), can be rather easily prevented. For example, in case of lung cancers, melanomas and cervical cancers some 85–100% of the cases can be avoided by retaining from smoking, avoiding UV exposures and vaccination against HPV, respectively. However, also **secondary prevention**, i.e., early detection and intervention, is saving lives, since not all types of cancers can be avoided by reducing environmental risk factors. For cancers that are primarily caused by random mutations due to uncorrected replication errors (Sect. 4.4), secondary prevention is even the major option. Prevention of metastasis combines local control of the cancer, such as surgery or radiotherapy (Sect. 11.1), with systemic therapy, such as inhibitors of overactive oncogenes (Sect. 11.2) and/or immunotherapy (Chap. 10). The latter targets cells that have left the primary malignant tumor and are not detectable by non-invasive imaging, i.e., they cannot be removed by surgery (Sect. 9.3).

Identifying and eliminating pre-cancerous lesions can prevent cancer, such as detecting polyps in colonoscopy and removing them. This decreases the risk for colorectal cancer by 75–90%. Similarly, the elimination of pre-cancerous cells from the cervix reduces the risk for respective cancer. Moreover, carriers of inherited mutations of the tumor suppressor genes *BRCA1* and *BRCA2* (Sect. 5.1) often decide on radical preventive mastectomy, i.e., complete resection of the breasts, while frequent cancer surveillance focusing on early detection and intervention of lesions in the breast may be sufficient.

There is a clear advance in early cancer detection technologies, but one cannot expect that they will keep our body completely free of pre-cancerous cells. **The best strategy is to engage everyone's immune system in the systematic elimination of transformed cells** (Chap. 10). One option is prophylactic vaccination against cancers that are caused by tumor viruses, such as HPV and HBV (Box 1.2). Another approach would be to reduce chronic inflammation due to obesity and other inflammation-related diseases, since inflammation significantly contributes to cancer onset (Sect. 8.4).

Box 1.2: Causes of cancer and potential reduction in cancer burden through preventive actions. There are factors that are proven to increase the risk of cancer:

- cigarette smoking/tobacco use
- exposure to radiation
- immunosuppression
- infections.

Obviously, avoiding these factors is the best way to prevent cancer (Table 1.1). For example, the great majority of lung adenocarcinoma cases (89%) are preventable. Moreover, **30% of all cancer deaths in the US are associated with cigarette smoking**. In addition, there is huge potential in the vaccination against viruses that are known to cause cancer, such as HPV and HBV. In the developing world, 1 in 4 cases of cancer originate from viruses or other infections, while in the developed world the rate is 1 in 10. In these countries, the contribution of the risk factors smoking, physical inactivity and obesity are significantly higher.

Cancer surviving children are due to aggressive treatment, such as radiotherapy, prone to secondary cancers in later years of their life. Therefore, some treatment protocols of pediatric cancers switched from the overall goal of curing to a compromise between curing and reducing the late side effects of the cancer treatment. For example, in the case of ALL "non inferior" treatment options avoid prophylactic brain radiation through the administration of the folic acid antagonist methotrexate.

Environment, diet and lifestyle affect the risk of cancer but studies, such as randomized controlled trials, measuring their impact, are difficult to conduct. Respective trials, such as investigating the impact of diet, require a long-term compliance of the participants. Recommendations for cancer prevention are:

- avoid risk factors
- preventive lifestyle with changes mainly in diet and physical activity
- early detection of pre-malignant lesions
- chemoprevention

- preventive surgery.

Early detection is well established for cervical, breast and colorectal cancer, but there is a risk of overdiagnosis. Chemoprevention are medicines that reduce an elevated cancer risk, e.g., of persons with cancer predisposition syndromes, by drugs aiming to interrupt tumorigenesis by activating bypass pathways. Finally, some individuals with cancer predisposition syndromes, such as carriers of mutations in *BRCA1* and *BRCA2*, choose for preventive surgery, i.e., organs like breast and ovaries are removed

More than 50% of today's cancer cases could be prevented by consistently using the cellular and molecular understanding on cancer onset. Thus, **cancer burden could be reduced significantly by changing the individual and population behaviors concerning tobacco smoking, obesity and physical inactivity**. Even about a third of cancers that are primarily based on mutations of driver genes are preventable, since they still linked to environmental factors. The concept of tumorigenesis (Sect. 1.4) implies that preventing one environment-triggered step in this multi-step process is in most cases sufficient in avoiding the life-threatening metastatic form of the cancer.

The main goal of cancer prevention is to inhibit that a cancer progresses to the final stage of invasive, potentially lethal metastatic cancer. Since the process of multi-step tumorigenesis takes time, prevention is the more effective the earlier in life it starts (Fig. 1.6). Cancer prevention either completely avoids the occurrence of a tumor of detectable size over lifetime or at least significantly delays its manifestation. Preventive interventions make even sense after the diagnosis of a malignant tumor, since they may slow down cancer progression. Interestingly, pre-cancerous lesions even have the potential to regress, which may look as being spontaneous but in reality represent the success of preventive measures that derive from the immune system. Thus, **activating the immune system,** e.g.**, by increased physical activity, is probably the most effective preventive measure in every stage of a cancer**.

There is no doubt that **primary cancer prevention is the most efficient way in reducing the global cancer burden**. Nevertheless, the vast majority of cancer research (and respective funding) is focused on late-stage metastatic cancers that are seldomly cured (Sect. 9.4). This focus has numerous reasons, such as the size of necessary clinical trials, regulatory affairs and the interest of pharma industry selling drugs focusing on late-stage cancer phases (Chap. 11). In other non-communicable diseases, such as cardiovascular disorders, significant effort is taken for preventing the disease by drugs lowering blood pressure or cholesterol level as well as on weight loss and increased physical activity, while **in cancer less comparable preventive achievements are undertaken**. We have to be realistic about that even the best new precise cancer therapies (Sect. 11.3) provide in most cases only transient responses, i.e., that they reduce the individual's disease burden only for a limited time. In

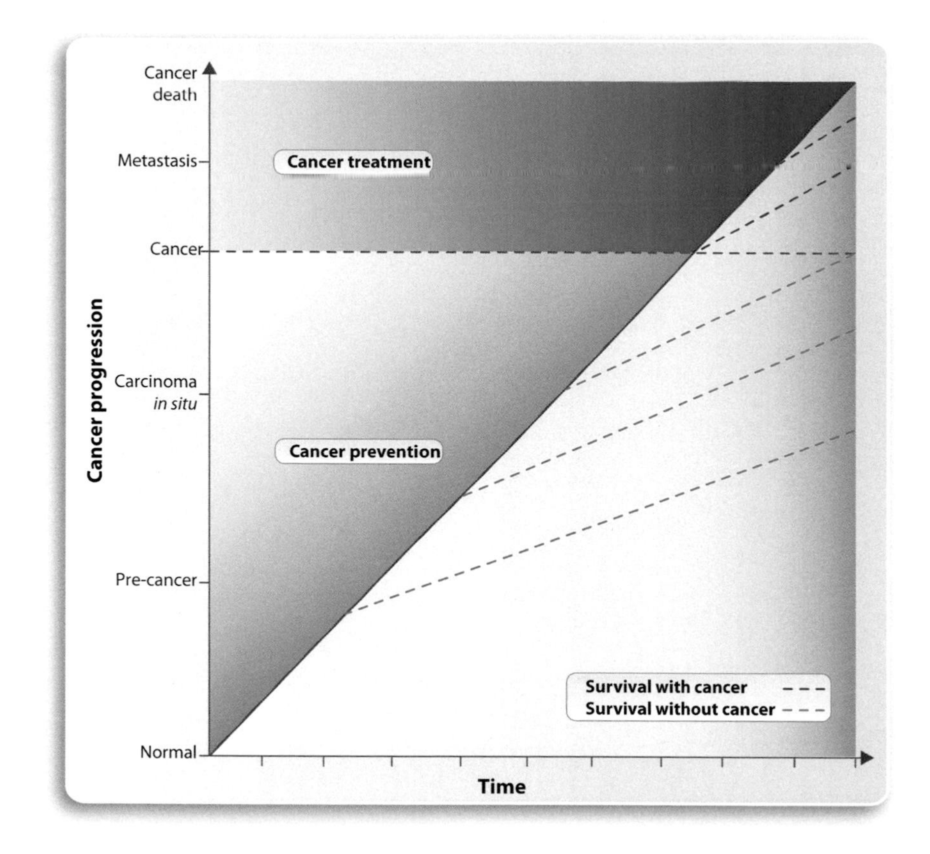

Fig. 1.6 The impact of cancer preventive interventions. The process of multi-step tumorigenesis (y-axis) suggest that the earlier in life (x-axis) one starts cancer preventive interventions, the more effective they are (indicated by a less steep slope). Ideally, the cancer-initiating process should be interrupted before pre-cancers form, but even after cancer diagnosis preventive interventions can delay disease progression and death

contrast, **a prevented cancer is cured and not only driven into a transient remission**. In this context also the avoidance of the mortality risk of surgery and adjuvant therapy needs to be considered.

Clinical conclusion: Cancer is a wide-spread disease affecting every second individual in industrialized countries. Many types of cancers are avoidable by lifestyle changes. Thus, primary cancer prevention would be the most effective way handling the global cancer burden.

Further Reading

Colditz, G. A., Wolin, K. Y., & Gehlert, S. (2012). Applying what we know to accelerate cancer prevention. *Science Translational Medicine, 4*, 127rv124.

Global Burden of Disease Consortium. (2018). Global, regional, and national age-sex-specific mortality for 282 causes of death in 195 countries and territories, 1980-2017: a systematic analysis for the Global Burden of Disease Study 2017. *Lancet, 392,* 1736–1788.

Rozenblatt-Rosen, O., Regev, A., Oberdoerffer, P., Nawy, T., Hupalowska, A., Rood, J. E., et al. (2020). The human tumor atlas network: Charting tumor transitions across space and time at single-cell resolution. *Cell, 181,* 236–249.

Siegel, R. L., Miller, K. D., & Jemal, A. (2020). Cancer statistics, 2020. *CA: A Cancer Journal for Clinicians, 70,* 7–30.

Thomas, F., Nesse, R. M., Gatenby, R., Gidoin, C., Renaud, F., Roche, B., et al. (2016). Evolutionary ecology of organs: A missing link in cancer development? *Trends Cancer, 2,* 409–415.

Tomasetti, C., Li, L., & Vogelstein, B. (2017). Stem cell divisions, somatic mutations, cancer etiology, and cancer prevention. *Science, 355,* 1330–1334.

Zahir, N., Sun, R., Gallahan, D., Gatenby, R. A., & Curtis, C. (2020). Characterizing the ecological and evolutionary dynamics of cancer. *Nature Genetics, 52,* 759–767.

Chapter 2
Oncogenes and Signal Transduction

Abstract The observation that RNA tumor viruses, which have captured host genes encoding for key proteins in signal transduction cascades, can transform normal cells into cancer cells led to the discovery of oncogenes. Oncogenes can be activated by point mutations, translocations and amplifications resulting in the uncontrolled and enhanced activity of signal transduction pathways. This disconnects the control of cellular growth, metabolism and survival from exogenous signals, such as growth factors, cytokines, hormones and neighboring cells, and leads to selective growth advantages of the cancer cells. Thus, the aberrant activation of oncogenes and their signaling networks allows transformed cells to acquire key hallmarks of cancer, such as evading growth suppressors, sustaining proliferative signaling and resisting cell death. The knowledge of tumor-specific genetic rearrangements is used in the diagnosis of cancer and in therapy monitoring but also holds prognostic information. Moreover, these molecular changes present therapeutic targets.

Keywords Cellular transformation · Tumor virus · Oncogene · Signal transduction · Point mutation · Amplification · Translocation · Hallmarks of cancer

2.1 Cellular Transformation

In normal human tissues, cells stop growing when organs have reached their final size. Similarly, cells arrest their cell cycle after they have replaced others that were dying, e.g., due to mechanical stress in skin or intestine. Most of these cells have then reached the status of terminal differentiation, i.e., they are in the G_0 phase of their cell cycle (Sect. 3.2). Similarly, also in the process of wound healing (Sect. 8.1) cells know, when they have to stop growing after destroyed cells have been replaced, i.e., when homeostasis is reconstituted. The control of proliferation and differentiation is mediated by the careful control of:

- anti-growth signals that are secreted by neighboring cells
- inhibitory signals due to physical contacts with the latter
- the lack of growth-stimulating signals.

C. Carlberg and E. Velleuer, *Cancer Biology: How Science Works*,
https://doi.org/10.1007/978-3-030-75699-4_2

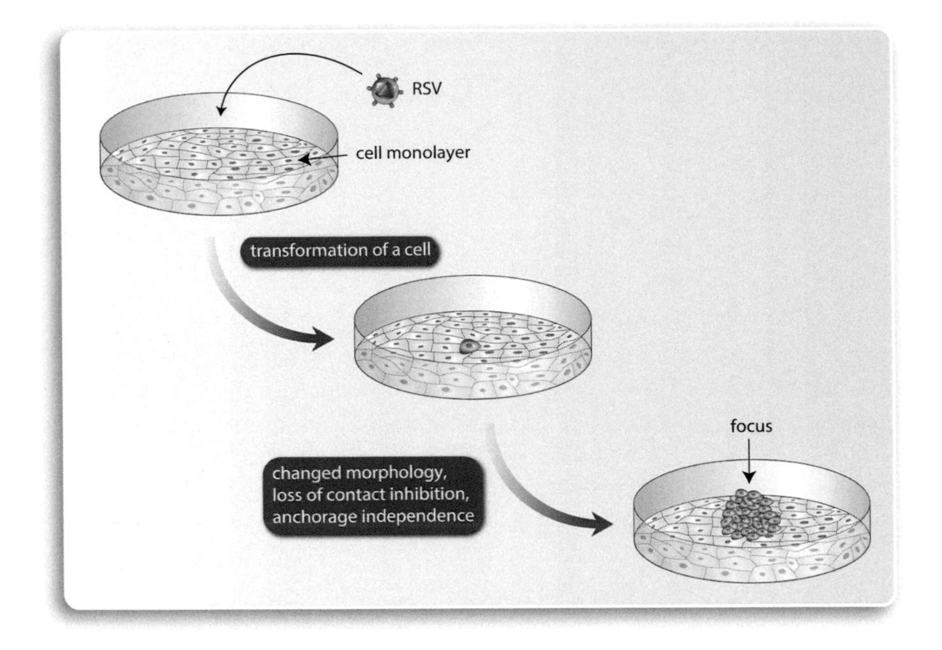

Fig. 2.1 Cellular transformation. This in vitro cell culture experiment illustrates the principle of cellular transformation (here induced by the RNA tumor virus RSV) of normal adherently growing cells into cancer cells showing lack of contact inhibition and forming a multi-layer clump known as a focus. Similar principles apply also for in vivo tumorigenesis

A fundamental trait of cancer cells is their uncontrolled growth. Thus, **the cellular basis of cancer onset is that one or multiple cells of our body transform into cells with advanced growth potential and reduced differentiation status**. In cell culture, adherently growing normal cells, e.g., fibroblasts, stop proliferating when the culture dish reached 100% confluency (Fig. 2.1, top). However, when one of these cells is transformed to a cancer cell, e.g., through an infection with a RNA tumor virus, such as Roux sarcoma virus (RSV) (Box 2.1), it loses the contact to its neighboring cells, changes its shape and forms a cell clone, a so-called focus, were identical cells grow on top of each other (Fig. 2.1, center and bottom).

Box 2.1: Tumor viruses. Most viruses do not cause cancer, but some of them, carrying either a **DNA or RNA genome**, are able to transform normal cells into cancer cells. Importantly, worldwide approximately 16% of all cancer cases are associated with a virus infection. DNA tumor viruses, such as Simian virus 40 (SV40), HPV and adenovirus, impair the function of the tumor suppressor proteins p53 and RB (Chap. 3), so that the viruses can replicate more efficiently. The RNA virus HBV can cause liver cancer, the human T lymphotropic virus is involved in adult T cell leukemia and HPV infection can cause cervix carcinoma, all of which could be prevented by vaccination. Some RNA tumor viruses are **retroviruses** that in addition to the genes *gag* (internal virion protein), *env* (envelope protein) and *pol* (reverse transcriptase) carry the information for a signaling protein, the origin of which is the host's genome. RSV was the first discovered RNA tumor virus, which carries a mutated version (*v-src*) of a normal cellular gene (a proto-oncogene) that encodes for the tyrosine kinase SRC (SRC proto-oncogene, non-receptor tyrosine kinase). RSV-infected cells express large amounts of intrinsically active SRC proteins and therefore rapidly transform into cancer cells (Fig. 2.1). Based on this and other viral oncogenes the **concept of oncogenes** has been developed (Sect. 2.2)

The major hallmarks of cellular transformation of adherently growing cells are:

- **changed morphology**, such as round shape
- **loss of contact inhibition**, i.e., the ability to grow on top of each other
- growth without attachment, i.e., **anchorage independence**
- indefinite growth, i.e., **immortalization**
- reduced requirement of mitogens, i.e., **independence of growth factors**.

This implies that due to transformation **cancer cells become masters of their own destinies**, i.e., they ignore anti-proliferative signals from their microenvironment and are not dependent on the presence of mitogens, such as growth factors.

The activation of a number of different molecular pathways can result in cellular transformation. The cells may produce themselves growth factors or stimulate neighboring normal cells to supply them with mitogens. Alternatively, the number of growth factor receptors may be upregulated or structurally changed by point mutations of their encoding genes, in order to make the transformed cells hypersensitive to limited amounts of growth factors. Furthermore, the constitutive activation of proteins further downstream in mitogenic signal transduction cascades, such as adaptor proteins, kinases and transcription factors, are sufficient for transforming normal cells into cancer cells (Sect. 2.2).

2.2 Activating Oncogenes in Signal Transduction Pathways

Principles discovered in the context of RNA tumor viruses define **oncogenes** as cellular genes encoding for key proteins in signal transduction cascades, i.e., for ligands, membrane receptors, intracellular adaptor proteins, kinases and transcription factors (Box 2.2). Most commonly oncogenes are activated by point mutations, which are either single nucleotide variants (SNVs), such as in the genes *RAS* (rat sarcoma) or *PIK3CA* (phosphatidylinositol-4,5-bisphosphate 3-kinase catalytic subunit alpha), or short insertions or deletions (indels). Other types of oncogene-activating mutations are amplifications or deletions, genomic rearrangements, such as translocations, and epigenetic activation, which will be discussed in Sects. 2.3 and 6.1, respectively.

Box 2.2: Signal transduction cascades. Signal transduction pathways have a common composition and typically start with an extracellular **ligand** binding to its cognate **membrane receptor**, which than at its cytosolic part interacts with **adaptor proteins** and activates them. This activation status is mostly transferred to a cascade of **protein kinases** that eventually either stimulate a **transcription factor** or a chromatin modifying enzyme (Sect. 6.1) in nucleus or other key enzymes located in the cytosol. In this way, extracellular (and in part also intracellular) signals are perceived by dozens of **signal transduction pathways**. The latter form networks of cellular circuits that represent key lines of intracellular communication, i.e., mechanisms by which a cell integrates and transduces intra- and extracellular signals from their source, e.g., the extracellular environment, into an action of the cell, such as to grow, differentiate or keep homeostasis, i.e., of its fate (Fig. 2.2A). Since proteins encoded by genes regulating cell fate, cell survival and genome maintenance often interact with each other, their pathways overlap, i.e., their classification may not be as distinct as indicated. Two pathways with key importance for cellular growth, RAS and PI3K, are illustrated (Fig. 2.2B). Aberrant activation of signal transduction pathways is one of the most frequent events in tumorigenesis, since it disconnects the control of cell growth, survival and metabolism from the control by mitogens, such as growth factors

One general principle of cancer is that the activation of **each of the few hundred oncogenes show the same result: they drive the process of tumorigenesis by providing transformed cells with selective growth advantages in comparison with neighboring normal cells** (Sect. 4.2). These cancer driver genes can be classified into a limited number (12, based on latest data 21 (Chap. 5)) of signal transduction cascades, which themselves can be sorted into three major cellular processes (Fig. 2.2A):

- **cell survival**: Transformed cancer cells are able to proliferate under limited nutrient conditions, which gives them selective growth advantages in comparison to normal cells in their microenvironment. For example, mutations in the oncogenes *KRAS* or *BRAF* (B-Raf proto-oncogene, serine/threonine kinase) allow cells to grow at lower glucose concentration than normal cells. Moreover, driver genes that directly regulate the cell cycle or apoptosis, such as *MYC* (MYC proto-oncogene, BHLH transcription factor) (Sect. 2.3) and *BCL2* (BCL2 apoptosis regulator), are often mutated in cancers. In addition, mutations of the gene *VHL* (Von Hippel-Lindau tumor suppressor) enhance cell survival and stimulate angiogenesis by the secretion of vascular endothelial growth factor (VEGF).

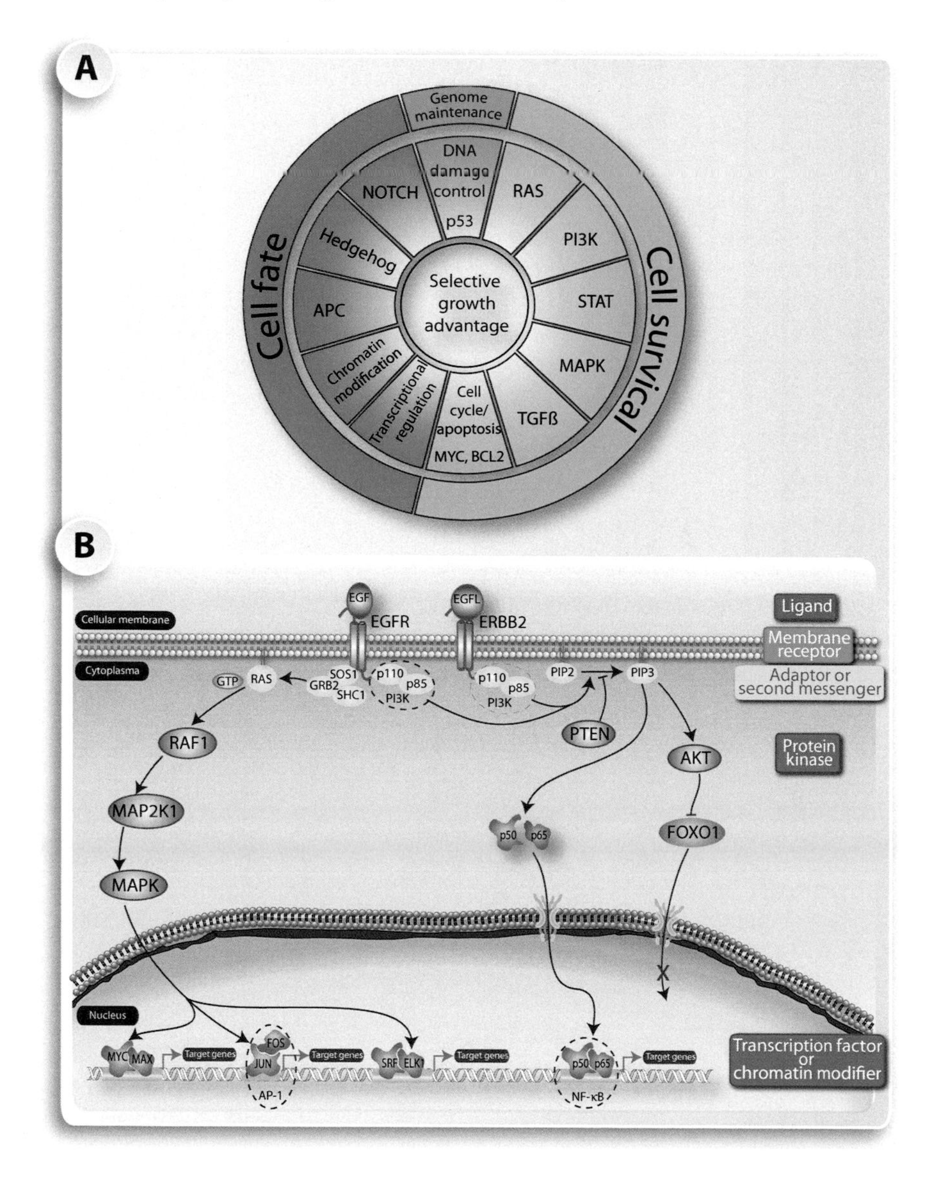

Fig. 2.2 Signal transduction pathways affected by mutations in cancer. Tumorigenesis is driven by the activation of oncogenes and the inactivation of tumor suppressor genes, which can be classified into one or more of signal transduction pathways (majorly organizing three core cellular processes, outer ring) (**A**). This confers transformed cells with a selective growth advantage (inner circle). The main components of two representative pathways (RAS and PI3K-AKT) are illustrated (**B**). Please note for this and following figures, the nuclear membrane is illustrated only with a single bilayer, while in reality it is a double bilayer. EGF = epidermal growth factor, EGFL = EGF like, ELK1 = ETS transcription factor ELK1, FOXO1 = forkhead box O1, GRB2 = growth factor receptor bound protein 2, MAP2K = mitogen-activated protein kinase kinase, MAPK = mitogen-activated protein kinase, NF-κB = nuclear factor κB, RAF1 = Raf-1 proto-oncogene, serine/threonine kinase, SHC1 = SHC adaptor protein 1, SOS1 = SOS Ras/Rac guanine nucleotide exchange factor 1, SRF, serum response factor, STAT = signal transducer and activator of transcription, TGFβ = transforming growth factor β

- **cell fate**: There is an opposing relationship between proliferation and differentiation, i.e., **proliferating cells are not terminally differentiated while terminally differentiated cells do not proliferate**. Many mutations in cancer genes, which primarily belong to the pathways APC, hedgehog or NOTCH (Notch receptor), disturb this balance between growth and differentiation by favoring the former. Since differentiated cells eventually become quiescent or die, this causes a selective growth advantage.
- **genome maintenance**: Normal cells use different pathways to detect and repair DNA defects occurring in each cell generation and keep in this way the rates of spontaneous mutations very low. There are a few molecular checkpoints for the integrity of the genome, which primarily involve the tumor suppressor protein p53 and its partner proteins (Sect. 3.1). When genome damage is detected, either the DNA repair machinery is activated (Sect. 4.3), the cell cycle is arrested (Sect. 3.2) or the cells are stimulated to commit suicide (apoptosis) (Sect. 3.1). When transformed cells do survive DNA damage, they obtain a selective growth advantage, often because their p53 signaling is harmed.

The three *RAS* genes (*HRAS*, *KRAS* and *NRAS*) belong to the most common oncogenes in human cancer, since mutations of them are found in 20-25% of all malignant tumors and even in up to 90% of pancreatic cancer cases. *HRAS* and *KRAS* were found first in retroviruses causing sarcoma in rats, while *NRAS* had been identified in human neuroblastoma. All three genes encode for small GTPases, which are cell membrane-attached cytosolic proteins that bind in their active state guanosine triphosphate (GTP) and in their inactive state guanosine diphosphate (GDP) (Fig. 2.2B). **Signaling pathways downstream of active RAS proteins control a large variety of cellular processes**, such as actin cytoskeletal integrity, apoptosis as well as cellular proliferation, differentiation, adhesion and migration. Many different human cancers are driven by point mutations of the *KRAS* gene, the most common of which result in the G12V amino acid exchange in the GTPase domain and in the Q61K exchange in the GTP hydrolysis domain, both of which lead to constitutively active proteins. Therefore, inhibitors of the RAS signal transduction cascade are intensively studied as targets for cancer treatment (Sect. 11.2).

Another important pathway is that of the kinases PI3K (phosphoinositide 3-kinase)-AKT (Akt murine thymoma viral oncogene homolog, also known as protein kinase B (PKB)) (Fig. 2.2B). This pathway is activated by a large variety of extracellular ligands, such as insulin, cytokines and growth factors, i.e., it responds to cytokine receptors, integrins, G protein-coupled receptors and receptor tyrosine kinases (RTKs). Examples of the latter are EGFR (epidermal growth factor receptor) and ERBB2 (Erb-B2 receptor tyrosine kinase 2, also called HER2), both of which are encoded by key oncogenes. **The PI3K-AKT signaling transduction cascade regulates both metabolic processes, such as glucose metabolism, biosynthesis of macromolecules and maintenance of redox balance, as well as cellular growth**. PI3Ks are dimeric protein complexes formed by a catalytic and a regulatory subunit. The gene *PIK3CA* encodes for the catalytic subunit p110α of PI3K and is the most frequently mutated single oncogene across all cancers (Chap. 5). Accordingly, growth factor-independent activation of the PI3K-AKT pathway via *PIK3CA* point mutations is one of the main cancer drivers.

Activation of PI3K leads to the production of the second messenger PIP3 (phosphatidylinositol-3,4,5-triphosphate), which is counteracted through dephosphorylation by the phosphatase PTEN (phosphatase and tensin homologue), i.e., **PI3K and PTEN form a negative feedback loop in the control of cellular growth**. Interestingly, PTEN protein is encoded by a key tumor suppressor gene (Chap. 3). Loss-of-function of the PTEN protein through mutations in its gene or inhibition of *PTEN* expression by promoter methylations amplify PI3K signaling and promote tumorigenesis in a variety of cancers. High PIP3 levels activate one of three isoforms of the kinase AKT that phosphorylate a diverse set of downstream substrates mediating the control of cellular growth, survival and metabolism. Interestingly, amplification and gain-of-function missense mutations in *AKT* genes also often occur in human cancers.

2.3 Oncogenic Translocations and Amplifications

In addition to point mutations, oncogenes can also be activated by genomic rearrangements, such as **translocations**. Most translocations occur in hematopoietic cancers, such as leukemias and lymphomas, or mesenchymal tumors, such as sarcomas, and are often cancer type-specific. Abnormal chromosomes, such as the so-called "Philadelphia chromosome" (a short version of chromosome 22 resulting from a translocation between chromosomes 9 and 22), are found in some cancer types, such as in CML. The translocation of the Philadelphia chromosome results in a fusion between the constitutively active gene *BCR* (BCR activator of RhoGEF and GTPase) and the oncogene *ABL1* (ABL proto-oncogene 1, non-receptor tyrosine kinase). In this way, the *ABL1* oncogene obtains high constitutive activity. Similarly, a translocation between chromosomes 15 and 17 in AML creates a fusion protein between the protein PML (PML nuclear body scaffold) and the oncogene product RARA (retinoic acid receptor alpha). In Burkitt lymphoma a translocation was identified, in which the oncogene *MYC* on chromosome 8 is fused with the gene *IGH* (immunoglobulin heavy locus) on chromosome 14. In this childhood cancer, chronic infections, e.g., with Epstein-Barr virus, lead to high antibody production and MYC protein overexpression specifically in B cells, which causes this special form of B cell lymphoma.

MYC is another oncogene being discovered based on a retrovirus. In total, there are three genes, *MYC*, *MYCN* and *MYCL*, encoding for MYC oncoproteins that function as transcription factors. *MYC* overexpression compared to normal tissues is the main mechanism how the oncoprotein acts as key driver of tumorigenesis. In this way, *MYC* overexpression by fusion with the highly active *IGH* gene is a special case applying for B cell lymphomas. In many other cancers the genomic region of one of the three *MYC* genes is amplified and the prognosis of these cancer cases is inversely proportional to the number of amplifications (Box 2.3, Fig. 2.3). Moreover, the MYC protein is the endpoint of a number of signal transduction cascades, such

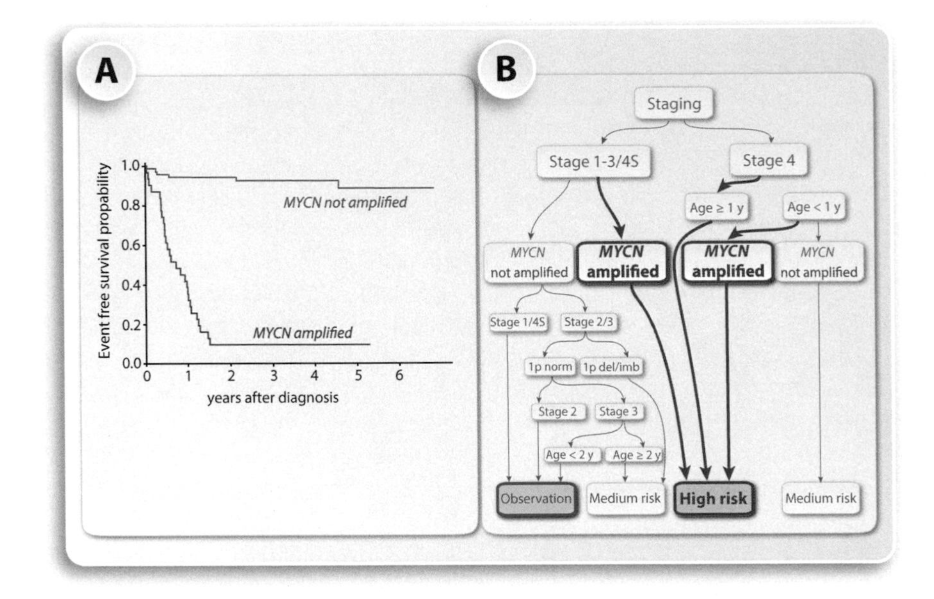

Fig. 2.3 Monitoring of *MYCN* overexpression and its effects on cancer survival. The *MYCN* expression status is of high impact for the prognosis (**A**) and treatment stratification (**B**) of patients with neuroblastoma. Patients with stage 4 neuroblastoma carrying *MYCN* amplification (red) have a far worse prognosis than those without *MYCN* amplification (green). Staging of pediatric neuroblastoma is outlined based on the German treatment protocol NB2004. Notably, *MYCN* amplification is the key criterion that is independent of other clinico-biological factors. Patients assigned to this high risk group receive the complete set of cancer treatment options, such as surgery, chemotherapy, radiotherapy including radiotherapy with I_{131}-metaiodobenzylguanidine (a norepinephrine analog that is specifically taken up by neuroblastoma cells), monoclonal antibody therapy and oral maintenance therapy with 13-*cis* retinoic acid. In addition, high dose chemotherapy is applied where the patient receives a bone marrow ablative dose, i.e., a drug concentration that is lethal for the bone marrow, followed by the rescue of the bone marrow via reinfusion of previously harvested patient's own hematopoietic stem cells. Data in **A** are based on Schmidt et al. (2000) J Clin Oncol *18*, 1260-1268

as the deregulated WNT pathway in colorectal cancer (Fig. 2.4). This leads to high levels of active MYC transcription factors in the respective malignant tumors.

Box 2.3: Clinical consequences of *MYC* overexpression. All three MYC proteins are oncogenic, but their tissue-specific expression pattern provides them with distinct characteristics in different types of cancer. MYCL is involved in small-cell lung carcinoma but is otherwise less characterized. MYC proteins are a valid target for cancer therapy, since many cancer cells depend on the enhanced expression of *MYC* gene, i.e., reversion of MYC overactivity would able to eradicate tumor growth. However, the ubiquitous expression of *MYC*, i.e., its involvement in many different processes makes the development of specific MYC inhibitors challenging. *MYCN* overexpression is one of the first tumor markers being identified already more than three decades ago in the context of pediatric neuroblastoma (Fig. 2.3). Neuroblastomas origin from the peripheral nerval system like the adrenal glands and are frequently detected with distant metastases due to the high growth rate

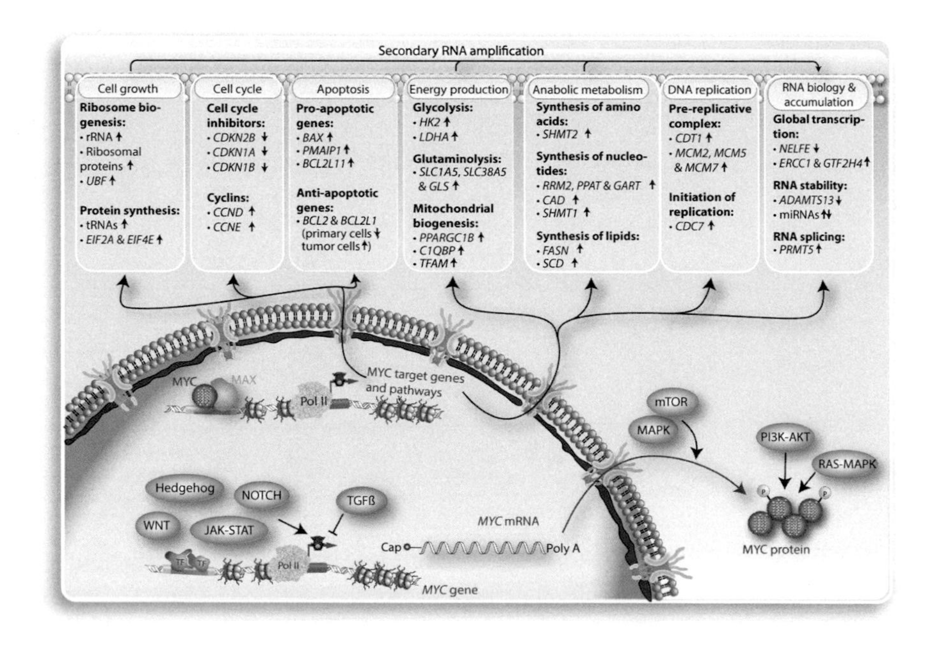

Fig. 2.4 The transcription factor MYC has a key position in cell growth-regulatory networks. The expression of the *MYC* gene is activated by major growth-regulatory pathways, such as NOTCH, WNT, JAK (Janus kinase)-STAT and Hedgehog, and inhibited by TGFβ signaling. Furthermore, the efficiency of *MYC* mRNA translation is affected by mTOR (mammalian target of rapamycin) and MAPK signaling, while the stability of MYC protein is enhanced by RAS and PI3K signaling. Examples of primary MYC target genes representing the indicated biological processes are listed. ADAMTS13 = ADAM metallopeptidase with thrombospondin type 1 motif 13, BAX = BCL2 associated X, apoptosis regulator, BCL2L = BCL2 like, C1QBP = complement C1q binding protein, CAD = carbamoyl-phosphate synthetase 2, CDC7 = cell division cycle 7, CDT1 = chromatin licensing and DNA replication factor 1, EIF = eukaryotic translation initiation factor, ERCC3 = ERCC excision repair 3, TFIIH core complex helicase subunit, GART = phosphoribosylglycinamide formyltransferase, FASN = fatty acid synthase, GLS = glutaminase, GTF2H4 = general transcription factor IIH subunit 4, HK2 = hexokinase 2, LDHA = lactate dehydrogenase A, MCM = minichromosome maintenance complex component, NELFE = negative elongation factor complex member E, PMAIP1 = phorbol-12-myristate-13-acetate-induced protein 1, PPARGC1B = PPARG coactivator 1 beta, PPAT = phosphoribosyl pyrophosphate amidotransferase, PRMT5 = protein arginine methyltransferase 5, RRM2 = ribonucleotide reductase regulatory subunit M2, SCD = stearoyl-CoA desaturase, SHMT = serine hydroxymethyltransferase, SLC = solute carrier family, TFAM = transcription factor A, mitochondrial, UBTF = upstream binding transcription factor

Like most other transcription factors, MYC binds as a dimeric complex to its DNA binding sites within enhancer and promoter regions of its target genes (Fig. 2.4). For gene activation MYC forms a heterodimer with the protein MAX (MYC associated factor X), while in repressing scenarios MYC interacts with the protein MNT (MAX network transcriptional repressor). Since MYC target genes are involved in nearly all anabolic and growth-promoting processes (Fig. 2.4), the overexpression of the oncoprotein provides cancer cells with selective growth advantages even in the absence of mitogenic signals. Importantly, sufficient *MYC* gene expression is essential during embryogenesis and MYC belongs together with OCT4 (octamer-binding transcription factor 4), KLF4 (Krüppel-like factor 4) and SOX2 (SRY-box 2) to a group of key proteins used for the induction of induced pluripotency stem (iPS) cells (Sect. 6.3). **A global overexpression of MYC affects not only its specific target genes but also enhances the overall expression rate of all active genes transcribed by RNA polymerase II** (Pol II), e.g., by shifting paused Pol II proteins into productive transcription elongation. This seems to be the central oncogenic function of MYC.

2.4 The Hallmarks of Cancer Concept

Basically, all types of cancer have acquired the same set of functional capabilities, referred to as hallmarks, that allow their cells to survive, proliferate, and disseminate. During the process of tumorigenesis, the order how the hallmarks are established as well as their mechanistic basis varies very much. The concept of the **hallmarks of cancer** aims to explain the cellular basis of the diversity of cancer types by focusing on the common properties of all cancer cells. Earlier sections of this Chapter showed that transformed cancer cells have a selective growth advantage that is primarily characterized by the hallmarks "sustained proliferative signaling" and "evading growth suppressors". Thus, in most types of cancer the initiating events are changes in the function and expression of cancer driver genes that provide transformed cells with one or both of these cellular properties. In the original formulation of the concept the four additional hallmarks "resisting cell death", "enabling replicative immortality", "inducing angiogenesis" and "activating invasion and metastasis" were listed. These six core hallmarks of cancer are cellular properties, all of which must be acquired, in order to transform stepwise a normal cell into an aggressive metastatic tumor cell (Fig. 2.5, left). Later on, the concept was extended by further hallmarks and enabling characteristics, including "genome instability and mutation", "deregulating cellular energetics", "avoiding immune destruction", "tumor-promoting inflammation" and "epigenomic disruption". This acknowledges the results of cancer genome projects concerning genetic and epigenetic drivers of different types of cancer (Chaps. 5 and 6).

A generalized model of tumorigenesis suggests that a normal cell acquires in a multi-step process all six core hallmarks, in order to transform into a metastatic cancer

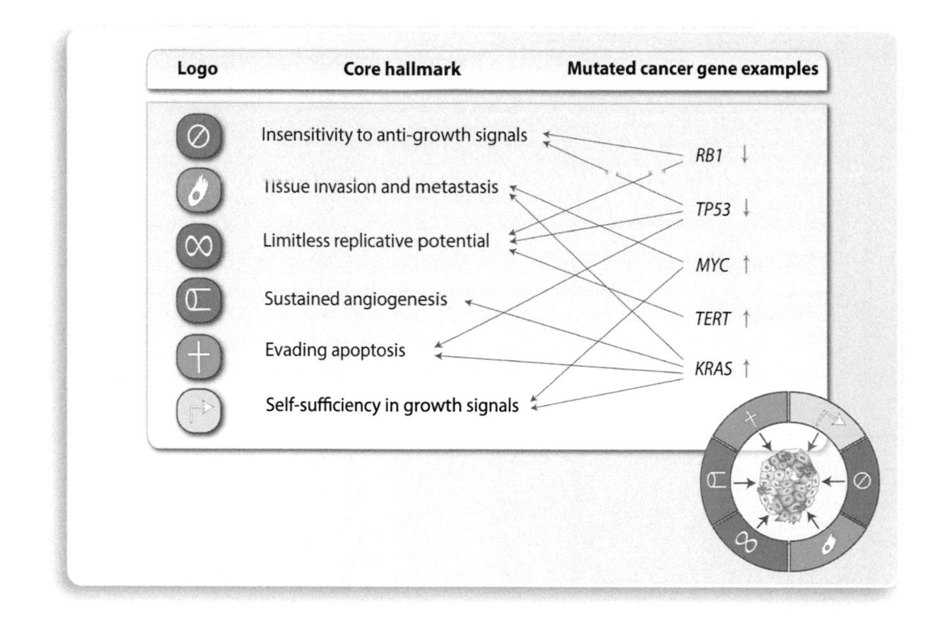

Fig. 2.5 Hallmarks of cancer. The six core hallmarks of cancer provide a logical framework for understanding the diversity of cancer types (**left**). The hallmarks can be acquired by mutations, genomic rearrangements and expression changes of the exemplified oncogenes and tumor suppressor genes (**right**)

cell (Sect. 4.2). The molecular basis of acquiring a specific hallmark is either an activating genetic or epigenetic change in a key oncogene, such as *KRAS* (Sect. 2.2), *MYC* (Sect. 2.3) or *TERT* (telomerase reverse transcriptase, Sect. 4.4), or a repressing alteration in a tumor suppressor gene, such as *TP53* (Sect. 3.1) or *RB1* (Sect. 3.2). Since the gene products of these oncogenes or tumor suppressor genes control multiple and diverse downstream targets of signaling pathways (RAS, PI3K, PTEN) or have a multitude of target genes (*MYC*, *TP53*, *RB1*), the (epi)genetic change of one cancer gene can affect more than one hallmark. For example, the activation of *RAS* genes or the inactivation of the *TP53* gene modulate four or three hallmarks, respectively, whereas the activation of the *TERT* gene or disabling the *RB1* gene affects only one or two hallmarks (Fig. 2.5, right). Accordingly, in order to acquire all key hallmarks as less as two driver mutations in key cancer genes are required, while in opposite case up to six mutational events in less important oncogenes or tumor suppressor genes would be necessary. Under the assumption that a given cell acquires every 10-20 years one driver mutation, this explains, at least in part, why cancer occurs in some persons already at rather young age while with other far later in life or never (Chap. 7). However, epimutations often occur more frequently than classical genetic mutations, i.e., **the larger proportion of a tumorigenesis process is driven by epigenetic changes, the earlier in life aggressive cancer may establish** (Sect. 6.4).

Clinical conclusion: On the cellular basis cancer is an hyperproliferative disease, in which pathways controling cellular fate are dysregulated. The molecular basis of this dysregulation is the activation of oncogenes encoding for key proteins in signal transduction cascades by point mutations, amplifications or translocations. Diagnosing these mutations is of increasing clinical importance for stratification of therapy, determining the individual's prognosis and developing targeted therapy.

Further Reading

Baluapuri, A., Wolf, E., & Eilers, M. (2020). Target gene-independent functions of MYC oncoproteins. *Nature Reviews Molecular Cell Biology, 21,* 255–267.

Hanahan, D., & Weinberg, R. A. (2011). Hallmarks of cancer: the next generation. *Cell, 144,* 646–674.

Hoxhaj, G., & Manning, B. D. (2020). The PI3K-AKT network at the interface of oncogenic signaling and cancer metabolism. *Nature Reviews Cancer, 20,* 74–88.

Vogelstein, B., Papadopoulos, N., Velculescu, V. E., Zhou, S., Diaz, L. A., Jr., & Kinzler, K. W. (2013). Cancer genome landscapes. *Science, 339,* 1546–1558.

Chapter 3
Tumor Suppressor Genes and Cell Fate Control

Abstract The transcription factor p53 is encoded by the most prominent cancer driver gene, the tumor suppressor gene *TP53*. Activated p53 proteins regulate hundreds of target genes that control cell fate controlling processes, such as DNA damage repair, cell cycle arrest, apoptosis and senescence. The aberrant activity of proteins ruling the cell cycle, such as the tumor suppressor protein RB and several cyclin-dependent kinase inhibitors (CDKIs) as well as oncogene-encoded cyclins and cyclin-dependent kinases (CDKs), leads to loss of proliferative control. The deletion of one or both alleles of a tumor suppressor gene (two hit model) or their gradual inactivation by transcriptional or epigenetic processes (continuum model) can explain both the inheritance as well as the mechanisms of cancer onset.

Keywords p53 · DNA damage response · Tumor suppressor gene · RB · CDKIs · Cell cycle control · Cyclins · CDKs · Tumor suppressor inactivation · Continuum model

3.1 p53 - A Master Example

The p53 protein is encoded by the tumor suppressor gene *TP53* (Box 3.1), the damage of which leads to severely reduced protection against cancer onset. The Li-Fraumeni syndrome is an inherited disease, the carriers of which have only one functional copy of the *TP53* gene and often develop cancers in early adulthood. Additionally, *TP53* is the most frequently mutated gene in human cancers, since more than 50% of malignant tumors contain a *TP53* mutation or deletion (Chap. 5). This explains why the *TP53* gene and its encoded protein are most studied of all times. Thus, p53 appears to be the key decision-making protein of our body that activates specific gene expression programs determining cellular outcomes, i.e., **p53 acts as a guardian of our genome**.

Box 3.1: Tumor suppressor genes. The homeostasis of normal cells is maintained by a physiological balance between the proteins encoded by tumor suppressor genes and oncogenes. The main characteristic of a tumor suppressor

C. Carlberg and E. Velleuer, *Cancer Biology: How Science Works*,
https://doi.org/10.1007/978-3-030-75699-4_3

protein is that it regulates pathways limiting inappropriate cell expansion, i.e., it inhibits cancer development and opposes oncogene function. Accordingly, the inactivation of tumor suppressor proteins by mutations within their genes or reduced expression facilitates tumor initiation or progression, i.e., **cells transform to cancer cells, when tumor suppressor genes are mutated or deleted**. The study of a rare inherited childhood tumor, retinoblastoma, led to the two hit model, based on which both alleles of a tumor suppressor gene, such as *RB1*, need to be inactivated, in order to cause a malignant tumor (Sect. 3.3). This explains the mechanisms behind a number of inherited cancers (Sect. 5.1). The final proof classifying a gene as a tumor suppressor is, when it impairs in an in vivo situation the onset or progression of a malignant tumor. For example, *TP53* knockout mice have a far higher cancer rate and die earlier than normal mice, in particular when they are challenged by mutagens or carcinogens

p53 was identified some 30 years ago based on its interaction with proteins, such as large T antigen and protein E6, encoded by the tumor viruses SV40 and HPV, respectively. The p53-inhibiting function of the viral proteins explains the tumor-promoting activity of the tumor viruses and highlights the tumor preventing potential of p53. The protein p53 is named by its apparent molecular weight detected by gel electrophoresis. Human p53 is composed of 393 amino acids that are subdivided into seven domains (Fig. 3.1). p53 is a unique transcription factor that binds DNA as a tetrameric complex and activates specific gene expression programs. The most severe mutations of *TP53* affect the oligomerization and DNA-binding domain (DBD) of the p53 protein, since they impair p53 DNA binding and cause the loss of its transcriptional activity.

p53 mediates activation as well as repression of its target genes, mostly via direct sequence-specific binding of the transcription factor to regulatory genomic regions, such as enhancers and promoters. Two copies of the RRRCWWGYYY (R = A or G, W = A or T, Y = C or T) binding motif, each contacting one p53 dimer, are separated by a spacer of up to 20 bp. Through protein-protein contacts p53 interacts with other transcription factors as well as with chromatin modifying enzymes (Sect. 6.1). Transcription factors, such as SP1 (specificity protein 1), CEBPA (CCAAT enhancer binding protein alpha) and AP-1 (activating protein 1, a heterodimer of the oncoproteins JUN (Jun proto-oncogene, AP-1 transcription factor subunit)) and FOS (Fos proto-oncogene, AP-1 transcription factor subunit)), are inactivated when they interact with p53. This is one mechanism how p53 acts as a tumor suppressor, i.e., it antagonizes the growth-promoting function of oncogene products, such as JUN and FOS.

p53 has a central role in the **DNA damage response (DDR)** process, but many other forms of cellular stress, such as hypoxia, telomere shortening, mitotic spindle damage, unfolded proteins, heat or cold shock, nutritional deprivation as well as improper ribosomal biogenesis, can induce p53 signaling. In general, reversible post-translational modifications of key amino acid residues within proteins are the

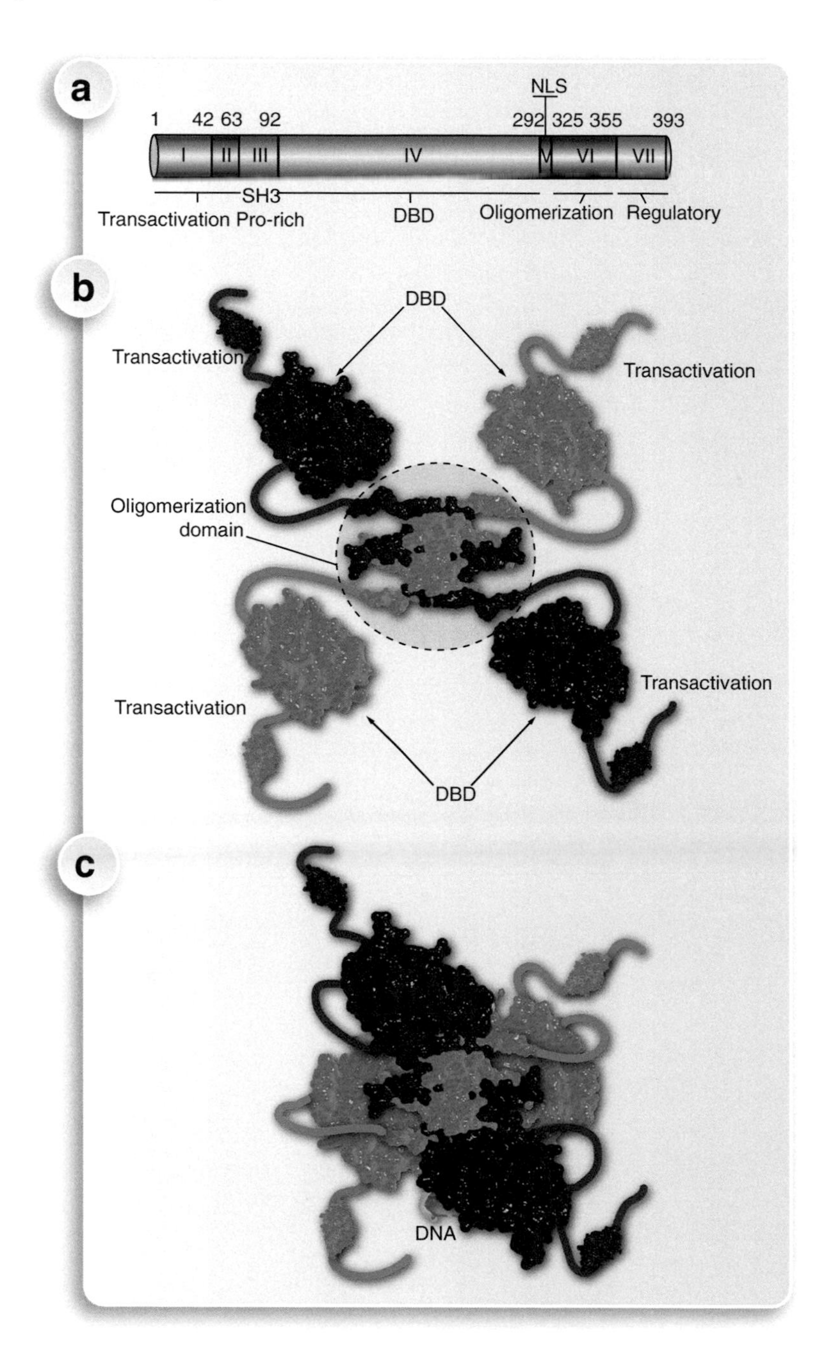

Fig. 3.1 Structure of p53. The human p53 protein is composed of seven subdomains: amino-terminal transactivation domain 1 (I), transactivation domain 2 (II), proline-rich domain important for the apoptotic activity (III), central DBD containing a zinc finger (IV), nuclear localization sequence (NLS) (V), oligomerization domain (VI) and carboxy-terminal domain important for downregulation of DNA binding (VII) (**A**). Model of the p53 tetramer in solution (**B**) and bound to DNA (**C**). Red and orange color refer to the two interacting p53 homodimers

major mechanisms of their communication and information storage for the control of signaling networks in cells. Also for the p53 protein post-translational modifications, such as phosphorylations, ubiquitinations, methylations and acetylations, have a huge impact, since they alter protein stability, DNA binding strength, target-gene selection and overall protein function. Accordingly, various proteins mediate information about cellular damage via post-translational modifications of p53 or its negative regulator MDM2 (MDM2 proto-oncogene, E3 ubiquitin protein ligase) (Fig. 3.2). MDM2 blocks the transcriptional activity of p53 via direct protein-protein interaction and leads to the degradation of the protein. In normal cells, p53 expression is rather low, but induced by stress signals MDM2 polyubiquitinates itself resulting in its degradation, which increases the half-life of the p53 protein from minutes to hours. Other stress sensing proteins are ATM (ATM serine/threonine kinase) and downstream of it PRKDC (protein kinase, DNA-activated, catalytic subunit) as well as ATR (ATR serine/threonine kinase) and its downstream target CSNK2A1 (casein

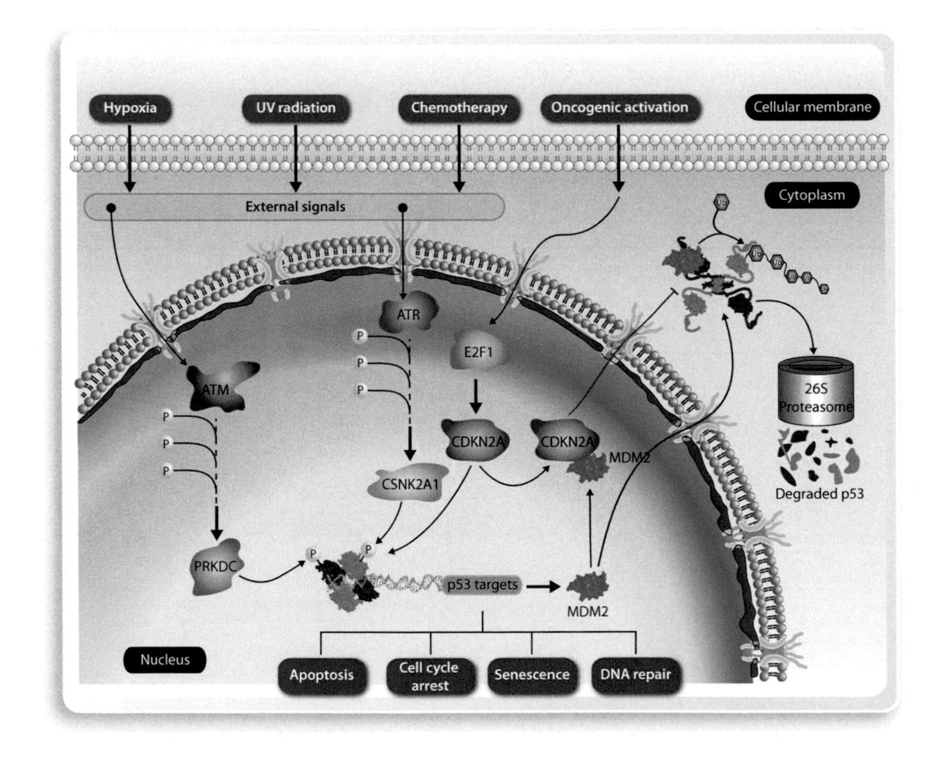

Fig. 3.2 The p53 pathway. When cells are stressed, signal mediator proteins are activated, which leads to phosphorylation of p53 or inhibition of p53 ubiquitination by MDM2. The half-life of p53 then increases from minutes to hours. The p53 tetramer recognizes its genomic binding sites controlling p53 target genes, one of which is *MDM2*. The tumor suppression function of p53 is mediated by genes controlling apoptosis, cell cycle arrest, senescence and DNA repair. More details are provided in the text

kinase 2 alpha 1) in addition to the transcription factor E2F1 regulating the cell cycle protein CDKN2A (cyclin-dependent kinase inhibitor 2A) (Sect. 3.2).

p53 is a central component of interconnected network of intracellular regulators and effectors. Depending on the interaction with different signal transduction pathways, activation of p53 leads either to DNA repair (Sect. 4.3), cell cycle arrest (Sect. 3.2), senescence (Box 7.1) or apoptosis (Box 3.2). Cell cycle arrest permits cellular repair, reverse of damage and cell survival, while senescence and apoptosis lead to cellular death. Interestingly, the p53 response is very flexible and depends on the cell type, its differentiation state, stress conditions and collaborating environmental signals. Thus, **the context, timing and extent of induction of p53 pathways have important implications for the fate of cells**.

Box 3.2: Apoptosis. In the context of malignant tumors, programmed cell death via apoptosis serves as a natural barrier to cancer development. During the course of tumorigenesis, apoptosis of cancer cells can be triggered by various physiologic stresses or anti-cancer therapy. Apoptosis-inducing stresses are signaling imbalances, e.g., they are based on elevated levels of oncogene signaling or DNA damage. Cancer cells use a number of strategies to limit or circumvent apoptosis, the most common of which is the loss of p53 function. Accordingly, "resisting cell death" is one of the core hallmarks of cancer (Sect. 2.4). On the molecular level, apoptosis is induced either by **extrinsic** factors, such as by Fas ligand and TNF, or by **intrinsic** proteins. Importantly, for the latter the counterbalance of pro- and anti-apoptotic proteins, many of which belong to the BCL2 family, is critical for the decision to continue with the apoptosis process via the release of the protein cytochrome C from mitochondria forming together with the protein APAF1 (apoptotic peptidase activating factor 1) and pro-caspase 9. Caspases form a cascade of latent proteases that execute the self-destroying of the cell induced both by the extrinsic and intrinsic apoptosis pathway. The progressively disassembled cell is consumed by neighboring cells and professional phagocytic cells in a silent way without stimulating an inflammatory response

3.2 Tumor Suppressors and Oncogenes in Cell Cycle Control

The cell cycle is a highly regulated process being essential for genome duplication and cell division. **Cancer is characterized by aberrant cell cycle activity** occurring either as a result of mutations in genes encoding for cell cycle proteins or for components of upstream signaling pathways. Cellular growth depends on the progression through the four distinct phases of the cell cycle: G_1, S, G_2 and M (Fig. 3.3). Most

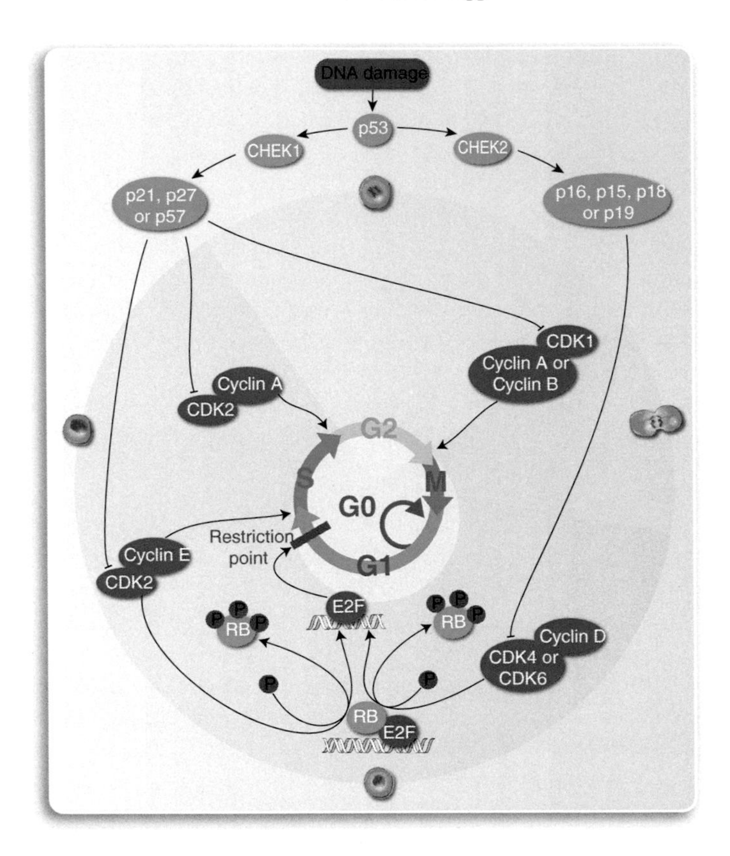

Fig. 3.3 Major regulatory proteins of cell cycle progression. Purple ovals indicate positive regulators of cell cycle progression, while blue ovals represent negative regulators. More details are provided in the text

cells of our body are terminally differentiated, i.e., based on programming of their epigenome and transcriptome they have arrested permanently their cell cycle and are considered to be in G_0 phase. In contrast, adult stem cells and progenitor cells keep their sensitivity to mitogenic signals or, in case of cellular transformation, normal cells regain it. Non-proliferating cells are in G_1 phase and carry a diploid genome. A progression from G_1 phase to S phase, in which genome duplicates, requires the activation of signal transduction pathways by mitogens, such as growth factors.

CDKs are the endpoint of mitogenic signal transduction cascades and many of them are encoded by oncogenes. However, the cell cycle controlling kinases are only active, when they form a complex with a cyclin protein. **Cyclins** are active only in a specific phase of the cell cycle, i.e., they determine in which phase which CDK is active. For example, CDK4-cyclin D and CDK6-cyclin D complexes are found in G_1 phase, CDK2-cyclin E at the begin of S phase, CDK2-cyclin A towards the end of S phase and CDK1-cyclin B towards the end of G_2 phase and begin of M phase. CDKs phosphorylate several cellular targets including the tumor suppressor protein RB. The latter acts as a repressor of transcription factors of the E2F family being critical for progression into S phase. In contrast to p53, RB integrates primarily signals of extracellular origin and "decides" whether or not the cell cycle should progress. Hyperphosphorylation of RB at the so-called **restriction point** (Fig. 3.3) releases its binding to E2Fs and allows the start of S phase. Important E2F target genes are *CCNE1* and *CCNE2*, which encode cyclins E1 and E2 activating CDK2. The cyclin E-CDK2 complex phosphorylates various drivers of cell cycle progression, DNA replication and centrosome duplication. The locus of the oncogene *CCNE1* is frequently amplified in ovarian cancer and breast cancer. Similarly, in hepatocellular carcinomas as well as in colorectal and breast cancers the oncogene *CCNA2* is often overexpressed via amplification.

In contrast, growth inhibitory signals antagonize G_1-S progression through the activation of **CDKI proteins** (Fig. 3.3). The CDKIs p16, p15, p18 and p19, which are encoded by the tumor suppressor genes *CDKN2A*, *CDKN2B*, *CDKN2C* and *CDKN2D*, respectively, bind to CDK4 and CDK6 and block their interaction with cyclins D1, D2 and D3. Genes encoding for proteins controlling the restriction point are commonly mutated in cancer. For example, the oncogene *CCND1* is located within the second most frequently amplified locus in human cancers, the oncogene *CDK4* carries in 50% of glioblastomas a constitutively activating point mutation and the oncogene *CDK6* is activated by translocations. Moreover, the tumor suppressor gene *CDKN2A* is located within the most frequently deleted locus in human cancers and is also often silenced via DNA methylation whilst the tumor suppressor gene *RB1* is deleted in many types of cancer (Chap. 5).

In G_1 phase **DNA damage** is sensed by several proteins activating the tumor suppressor protein p53 (Sect. 3.1) and triggers cell cycle arrest via checkpoint kinase 2 (CHEK2) or via the kinase CHEK1 in phases S or G_2 (Fig. 3.3). The CDKIs p21, p27 and p57 are encoded by the tumor suppressor genes *CDKN1A*, *CDKN1B* and *CDKN1C*, respectively, and inhibit CDK1 and CDK2.

Taken together, **oncogene-encoded cell cycle proteins are often overactive in malignant tumor cells**, which causes uncontrolled proliferation. This makes them promising targets for anti-cancer therapy, such as by CDK4/6-selective inhibitors (Sect. 11.2).

3.3 Tumor Suppressor Inhibition and Cancer Onset

The mechanisms of inactivation and deletion of tumor suppressor genes provide a rational for understanding hereditary predisposition to cancer. The **two hit model** (Box 3.1) suggests that both alleles of a tumor suppressor gene need to be lost or inactivated, in order to cause cancer. A classic example is the *RB1* tumor suppressor gene, the loss of one allele induces cancer susceptibility, while only after deletion or complete inactivation of both alleles the rare cancer type retinoblastoma occurs, often already in early childhood (Fig. 3.4, left). Increased cancer susceptibility means that

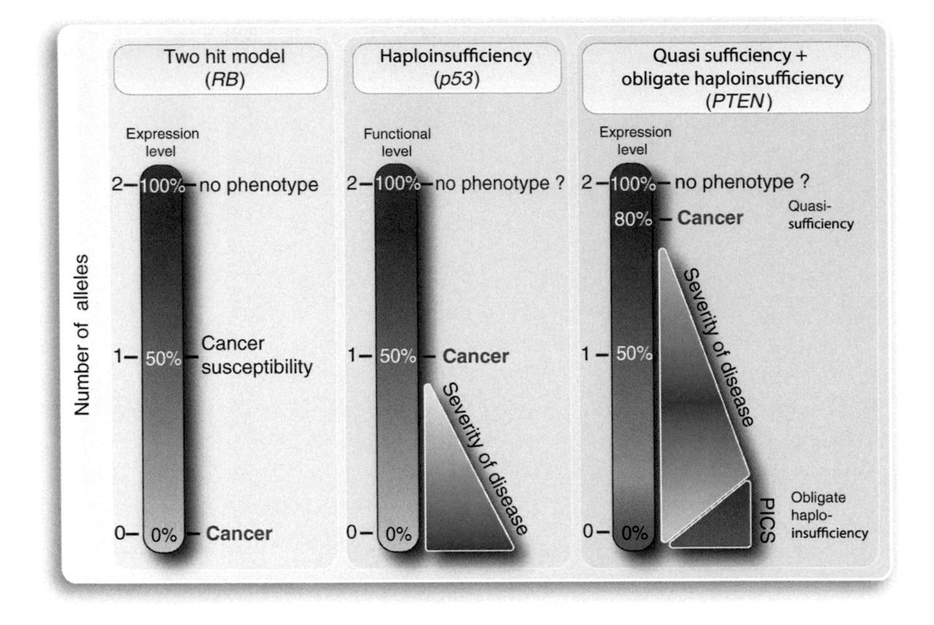

Fig. 3.4 Consequences of tumor suppressor loss. The examples of the tumor suppressor genes *RB1* (**left**), *TP53* (**center**) and *PTEN* (**right**) illustrate the consequences of the deletion of one or two alleles or their gradual inactivation (y-axis). The effect of complete loss of *PTEN* is highly context-dependent, since obligate haploinsufficiency can be caused by PTEN loss-induced cellular senescence (PICS)

the first hit, i.e., the inactivating mutation or deletion of the first allele that is inherited and present in the germline (i.e., in every cell of the body), significantly enhances the chance of obtaining within lifetime a second hit and developing the respective cancer type. Thus, in comparison of inherited versus sporadic cases of retinoblastoma the two hit model can explain differences in malignant tumor number and age of cancer onset. Moreover, the model explains a large part of cancer heritability, such as observed with mutations in the tumor suppressor genes *APC*, *BRCA1* and *BRCA2* (Sect. 5.1). Although the mutations in both alleles of a tumor suppressor gene are probably the cancer initiating events, they should not be misinterpreted as being sufficient for the development of a malignant tumor. As discussed in the context of hallmarks of cancer (Sect. 2.4), additional mutations in other cancer driver genes are necessary.

There are cases where the loss of one copy of a tumor suppressor gene is sufficient for inducing cancer, i.e., the mutations of these genes behave as being dominant negative. This phenomenon is referred to as **haploinsufficiency** and applies, e.g., for the *TP53* gene (Fig. 3.4, center). Since p53 functions as a tetrameric transcription factor (Sect. 3.1), one or two mutated copies of the protein can dominate the remaining normal copies and may significantly harm the functionality of the whole complex. When the second normal copy of the tumor suppressor gene may be reduced in expression, e.g., by promoter methylation, the mutated protein becomes even more dominant concerning the reduction of the complex functionality and the disease gets more severe. The effect of tumor suppressor gene haploinsufficiency depends on tissue and context, such as different thresholds of protein expression or the presence of compensatory proteins.

The tumor suppressor gene *PTEN* serves as example for quasisufficiency combined with obligate haploinsufficiency (Fig. 3.4, right). Quasisufficiency refers to the observation that a minor reduction in tumor suppressor gene expression, e.g., by 20%, can majorly harm the functionality of the respective tumor suppression pathway, so that cancer occurs. This effect is very tissue- and context-specific and depends, e.g., on the co-expression of other tumor suppressor genes, such as *TP53*. Also in this case the severity of the disease increases the more the expression of the tumor suppressor gene is diminished. Obligate haploinsufficiency is observed when a haploinsufficiency of the tumor suppressor gene would be more tumorigenic than a complete loss of the gene.

Taken together, there are two major models of tumorigenesis:

- for many tumor suppressor genes there are discrete states like in the two hit model with complete loss of both copies of the gene or in haploinsufficiency with a single gene loss (Fig. 3.5, left)
- for some tumor suppressor genes a continuum of expression level exists that result in a gradient of loss-of-function and increase of malignancy (Fig. 3.5, center).

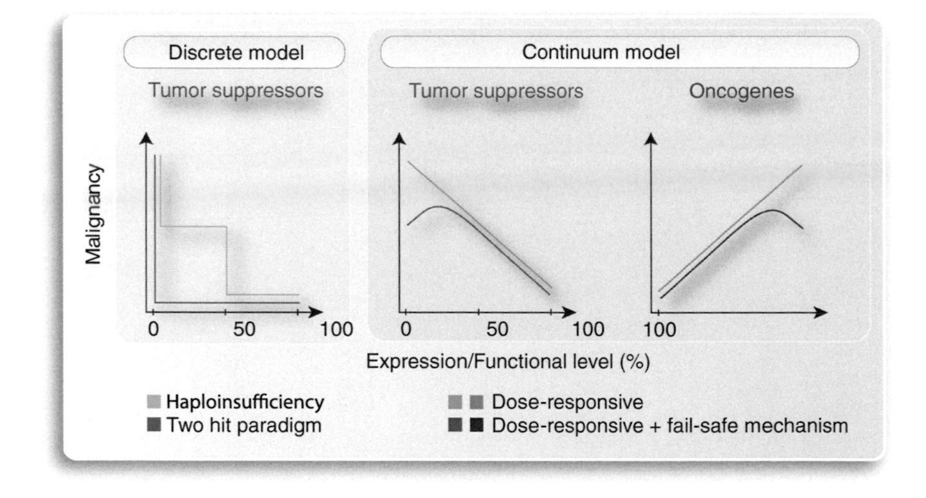

Fig. 3.5 Models of cancer gene activity. The classical discrete, step-wise model of tumor suppression (**left**) is compared with the continuum model of tumor suppression (**center**) and oncogenesis (**right**). The linear relationship in the continuum model is just a simplified assumption and may in reality be more complex. Moreover, failsafe mechanisms (darker lines), such as the induction of senescence, may be induced by complete loss of the tumor suppressor gene or massive overexpression of the oncogene, which are negatively correlated with malignancy

Importantly, the continuum model can also be applied for gradual activation of oncogenes and the respective boost in tumor growth (Fig. 3.5, right).

The continuum model suggests that some tumor suppressor genes, such as *PTEN*, are very sensitive to dose, others like *TP53* are moderately susceptibility to expression changes, while other tumor suppressor genes, such as *RB1* or *APC*, follow the two hit model. Furthermore, the continuum model implies that already minor changes in the epigenetic and transcriptional state of a tumor suppressor gene or the activity of the respective protein can have profound consequences on the susceptibility to cancer and its progression. Thus, **precise dosage of tumor suppressor gene expression is important for their proper function** (Fig. 3.6). The major regulatory mechanisms are:

- activation of promoter and enhancer regions through binding of specific transcription factors
- epigenetic regulation of these genomic regions via DNA methylation and post-translational histone modifications (Chap. 6)

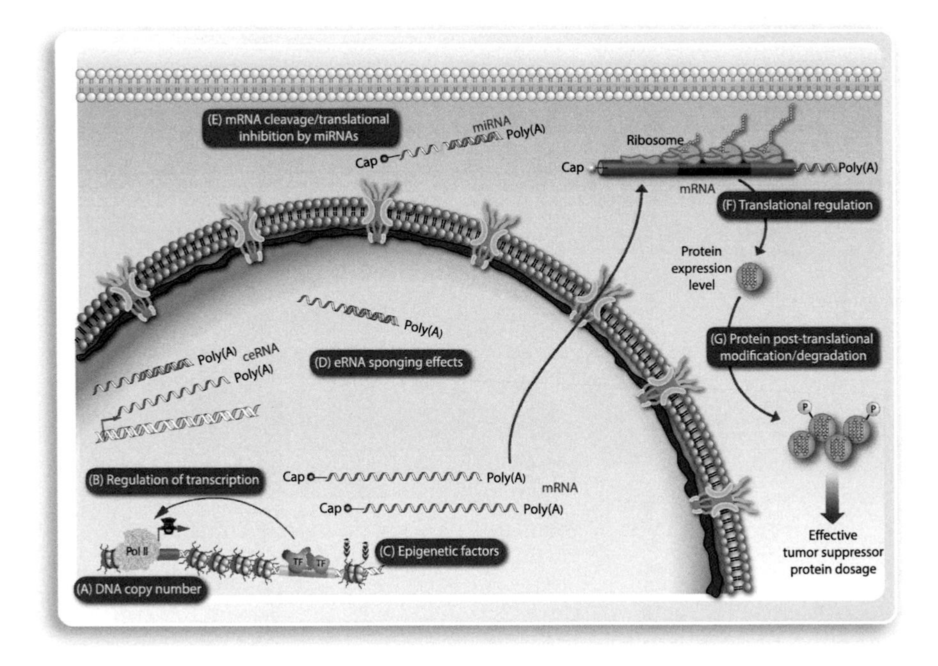

Fig. 3.6 Tumor suppressor gene dosage. The expression of a tumor suppressor gene and the functionality of the encoded protein depends on a number of mechanisms. These are the copy number of the gene (**A**), its transcriptional regulation by transcription factors (**B**) as well as the amount of its epigenetic silencing via DNA methylation (**C**). The mRNA level of tumor suppressor genes is negatively regulated by eRNAs (**D**) and miRNAs (**E**). Finally, also tumor suppressor protein translation (**F**) and post-translational modifications (**G**) contribute to expression level und functionality of the tumor suppressor protein

- fine-tuning of mRNA levels by ncRNAs, such as enhancer RNAs (eRNAs) and micro RNAs (miRNAs) (Box 3.3)
- affecting protein expression levels via the regulation of translation efficiency
- regulating protein functionality by post-translational modifications.

For example, many *PTEN* mRNA targeting miRNAs are amplified or overexpressed in different cancer types, such as miR-26 in glioma and miR-22 in prostate cancer.

Box 3.3: ncRNAs. Tens of thousands of RNA transcripts were discovered in human tissues and cell types that resemble mRNAs but do not translate into proteins, i.e., they are ncRNAs. The roles of ncRNA genes are quite diverse, including gene regulation, such as by miRNAs, RNA processing in splicing like by small nucleolar RNAs (snoRNAs) and protein synthesis like by transfer RNAs (tRNAs) and ribosomal RNAs (rRNAs). miRNAs are regulating the stability of mRNA molecules for a timespan as their translation products are needed in the cell. The names/identifiers in the database miRBase (www.mirbase.org) and in the literature are given in the hsa-mir-121 form, where the first three letters signify the organism, as in this case "hsa" for *homo sapiens.* Then, the mature miRNA is designated as "miR-121" (with capital R) in the database and in much of the literature, whilst "mir-121" (with small form r) refers to the miRNA gene and also to the predicted stem-loop portion of the primary transcript. Long ncRNAs are heterogeneous in their biogenesis, abundance and stability, and they differ in the mechanism of action. Some long ncRNA have a clear function, such as in regulation of gene expression and controlling chromatin accessibility, while others, such as eRNAs, may be primarily side products non-precise of Pol II transcription

The continuous nature of tumor suppressive gene expression is a key point in understanding interindividual variations in cancer susceptibility based on single nucleotide variations (SNPs) at the loci of cancer genes (Sect. 5.1).

Clinical conclusion: Tumor suppressor genes belong to the most important genes within our genome. They encode for proteins controlling cell cycle progression, DNA repair and the induction of senescence and apoptosis. Mutations or deletions of these genes explain not only the molecular basis of sporadic cancers but also inherited cancers. Knowing these mutations is of critical clinical importance for cancer surveillance (screening, preventive therapy) of the individual as well as for their family members.

Further Reading

Hafner, A., Bulyk, M. L., Jambhekar, A., & Lahav, G. (2019). The multiple mechanisms that regulate p53 activity and cell fate. *Nature Reviews Molecular Cell Biology, 20,* 199–210.

Kastenhuber, E. R., & Lowe, S. W. (2017). Putting p53 in context. *Cell, 170,* 1062–1078.

Levine, A. J. (2020). p53: 800 million years of evolution and 40 years of discovery. *Nature Reviews Cancer, 20,* 471–480.

Otto, T., & Sicinski, P. (2017). Cell cycle proteins as promising targets in cancer therapy. *Nature Reviews Cancer, 17,* 93–115.

Chapter 4
Multi-step Tumorigenesis and Genome Instability

Abstract The characterization of the growth of primary and metastatic tumors is crucial for the diagnosis and therapy of cancer. The process of tumorigenesis comprises multiple steps of mutations in cancer driver genes that provide the cell clone with a selective growth advantage over its neighboring cells. Importantly, the individual nature of each cancer case is based on two to six driver mutations out of a choice of some 500 oncogenes and tumor suppressor genes as well as on a flexible order how the hallmarks of cancer are obtained. The extraordinary ability of the different DNA repair pathways to maintain tens of thousands of DNA lesions, which each cell of our body is daily exposed to, is the reason that only approximately 40 mutations accumulate per cell and year. The vast amount of these mutations are passengers and only a few are drivers. The lifetime risk of many types of cancers correlates well with the number of stem cell divisions used for maintenance of tissue homeostasis. Accordingly, only a third of the cancer risk would be based on environmental exposure and inherited predisposition, while the majority has a stochastic basis.

Keywords Tumor growth · Metastatic tumors · Genome instability · Mutations · DNA repair pathways · Cancer risk · Stem cell division

4.1 Characteristics of Tumor Growth

More than 90% of the mortality to cancer is due to metastatic disease, i.e., when a primary malignant tumor has spread to a large number of mostly rapidly growing cancers in different tissues (Chap. 9). The likelihood of forming metastasis is generally increasing with the size of the primary malignant tumor. Unfortunately, the detection limit for a solid tumor by imaging methods (Box 4.1) is 10 mm in diameter. At this size **the tumor contains a billion (10^9) cells and may have undergone some 30 rounds of cell division** since cancer initiation by the transformation of a normal cell (Sect. 2.1) (Fig. 4.1A). After diagnosis the tumor is considered to be in its "visible phase" (Fig. 4.1B), in which it initially may grow exponentially but then slows down at larger sizes. In the previous "invisible phase" the tumor may have had different growth rates and accordingly 2–12 years may have passed since

C. Carlberg and E. Velleuer, *Cancer Biology: How Science Works*,
https://doi.org/10.1007/978-3-030-75699-4_4

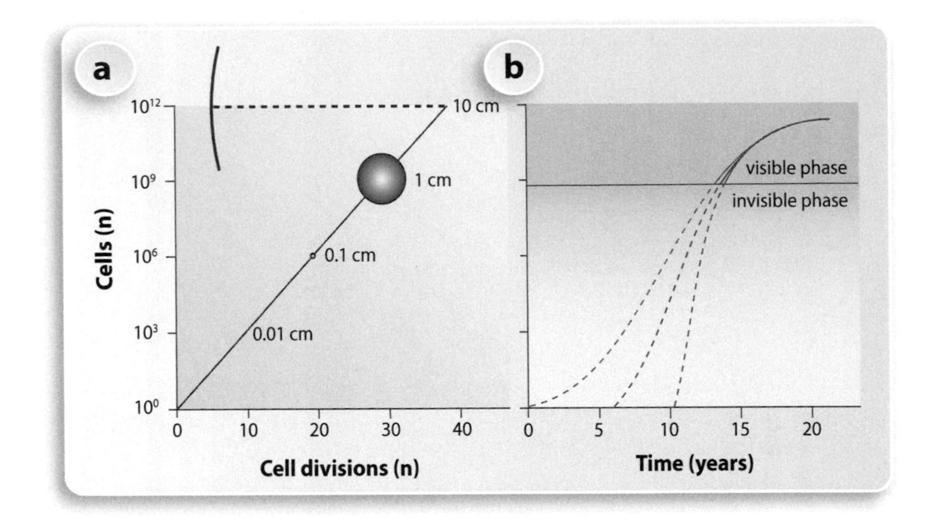

Fig. 4.1 Growth rates of malignant tumors. The growth rate of malignant tumors is exponential, so that the tumor cell number are displayed in logarithmic scale (y-axis) in relation to number of cell divisions and tumor diameter (**A**) and time since cancer initiation (**B**)

tumor initiation. **It is possible that the malignant tumor has already spread in this invisible phase**.

Box 4.1: Imaging methods for tumor detection. There are five major non-invasive imaging methods for the detection of solid tumors

- X-ray as plain film and as computed tomography (CT)
- ultrasound
- magnetic resonance imaging (MRI)
- radiation-assisted imaging like positron emission tomography (PET), single-photon emission computed tomography or bone scintigraphy
- optical imaging either by naked eye or via a microscope.

Each method has its advantages and disadvantages. X-ray and radiation-assisted imaging imply radiation exposure of the patient. Ultrasound has the highest anatomical resolution, i.e., it can detect tumors at already small sizes (<1 cm), but is highly dependent on the experience of the examiner. For MRI the patient has to lie a longer time without movement, so that in particular for children sedation is needed. For optical imaging, the site of interest needs to be accessible, which is not easily applicable for inner organs.

For liquid tumors, such as in leukemia, there are three main detection methods:

- peripheral blood or bone marrow smears

- fluorescence-activated cell sorting (FACS)
- molecular methods detecting the cancer clone (mostly via RT-PCR)

So-called liquid biopsies aim to detect tumor markers of solid cancers, such as fragments of tumor DNA in blood, urine or liquor. Furthermore, tests for blood in the stool is frequently used in screening for colorectal cancer, whereas the detection of catecholamines in urine (e.g., in neuroblastoma) can be applied exclusively during cancer detection.

Cancer detecting methods are primarily applied when a patient presents symptoms, but they can also be used for screening, i.e., for prevention (Sect. 1.5). However, for screening the methods have to be non-invasive and very accurate, in order to avoid overdiagnosis. Thus, **early detection of cancer ideally is a combination of a good own body feeling (any symptoms) and the appropriate screening and detection methods**

When during the process of tumorigenesis cells of a primary malignant tumor accumulate a new driver mutation, they are selected for competitive fitness (Fig. 4.2). After a few rounds of this selection process the cells are able to grow autonomously at a high rate. The more these malignant tumor cell clones expand, i.e., the larger size the primary malignant tumor growths, the more likely some cells leave the primary site and disseminate in the body and seed secondary malignant tumors. Thus, **the development of fully malignant cancer cells correlates with tumor size**. When the primary metastases have adapted to their distant sites and grow to a substantial size, some of their cells may propagate to a secondary metastasis with

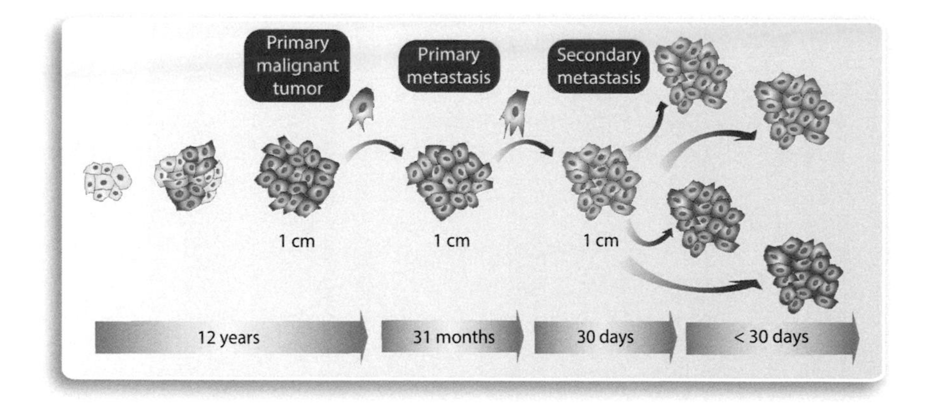

Fig. 4.2 The metastatic cascade model. In the process of tumorigenesis multiple rounds of mutation and selection for competitive fitness lead to cells that are able to leave the primary malignant tumor. In the example shown here this happens at a stage of 1 cm tumor diameter and may have taken 12 years. In primary, secondary and tertiary metastasis the cells become increasingly malignant and the tumor doubling time significantly decreases

even increased growth potential. Depending on tumor classification (Box 4.2) the disease may progress until the lethal mass of all metastases of approximately 1 kg in total, which is equal to more than 10^{12} cells.

Box 4.2: Tumor classification. The TNM classification represents the extent of disease at diagnosis by measuring tumor size (T stage) and cancer spread to lymph nodes (N stage) or metastatic sites (M stage). Tis indicates a carcinoma in situ, i.e., no invasion of the basement membrane. T1 is a small cancer and T4 a big cancer. N0 is defined by no spread of the cancer to a lymph node. According to the distance to the primary cancer there are three different subtypes (N1-N3) of lymph node invasion. In M0 no distant metastasis are found whereas in M1 this is the case. X means, that tumor size (Tx), lymph node invasion (Nx) or the status of metastasis (Mx) cannot be determined. A cancer stage (mostly 1-4) is composed of a combination of the above mentioned criteria (e.g., stage 1 breast cancer is T1N0M0) and is unique for each cancer subtype (Fig. 2.3B). Additionally to the stage, the grading (G1-G4) defines the grad of differentiation (G1: well differentiated, G4: anaplastic, i.e., highly **un**differentiated). Survival time is determined by the speed needed to reach a total tumor mass of about 1 kg. For example, the 15 year survival prognosis for patients with T1N0M0 breast cancer (tumor < 2 cm, no lymph node and no overt distant metastasis) is 90%, whereas for patients with T2N0M0 breast cancer (tumor diameter 2-5 cm, no lymph node and no overt distant metastasis) it is 70%

4.2 Multi-step Tumorigenesis

Tumorigenesis follows **evolutionary principles** of variation and selection. Spontaneously arising somatic mutations introduce genetic differences, i.e., variation between the cells of a tissue. Cells carrying a mutation that provides selective advantages over neighboring cells are positively selected leading to **clonal expansion** of the mutated cells. For example, when an epithelial cell obtains its first driver mutation (Sect. 4.4), it transforms to a cell with a **selective growth advantage** (Fig. 4.3, top left). This leads to a mild dysplasia, which via an additional driver mutations can develop to moderate dysplasia and later to severe dysplasia. The latter becomes so large that its further growth depends on angiogenesis (Sect. 8.3) and is characterized by the recruitment of immune cells, such as tumor-associated macrophages (TAMs) (Sect. 8.4). TAMs create a chronic inflammatory milieu that further promotes tumor development to a fully malignant form, some cells of which start evasion to other tissues (Fig. 4.3, top right). In colon cells a first mutation may inactivate the tumor suppressor gene *APC* and a small adenoma develops. This slow growing small

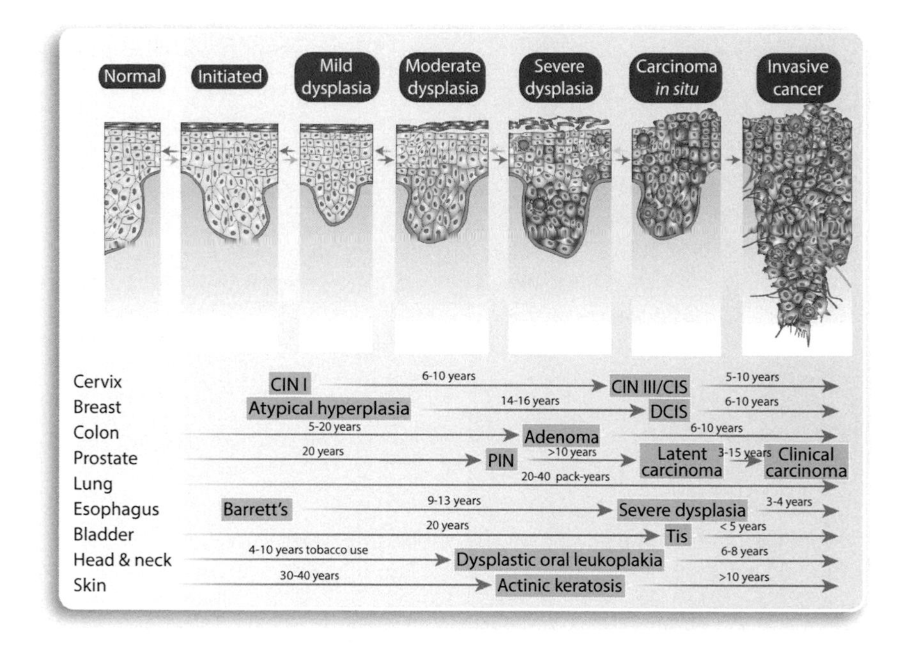

Fig. 4.3 Tumorigenesis. Tumorigenesis is a multi-step process that is portrayed as a succession of clonal expansions. Estimated timelines of the progression of selected cancers are indicated. CIN = cervical intraepithelial neoplasia, CIS = carcinoma in situ, DCIS = ductal carcinoma in situ, PIN = prostatic intraepithelial neoplasia, Tis = carcinoma in situ

adenoma then obtains in one of its cells a mutation in another driver gene, e.g., activating the oncogene *KRAS*. This allows the mutated cell to start a second round of clonal growth, i.e., it will overgrow neighboring tumor cells. In the vast majority of colorectal cancer cases driver mutations affecting at least four different signal transduction pathways are required for the progression to an invasive carcinoma. Thus, a few further rounds of mutations and clonal grow create an aggressive colon adenocarcinoma that may disseminate cells to other organs.

Somatic mutations in malignant tumor cell clones can be induced both by:

- **endogenous processes**, such as exposure to chemicals like in tobacco smoke, alcohol, aflatoxin or chemotherapeutic agents (Sect. 11.1), ionizing or UV radiation
- **cell-intrinsic processes**, such as DNA replication errors, impaired DNA repair (Sect. 4.3) or reactive oxygen species (ROS).

Exogeneous processes largely depend on lifestyle and can vary over time, while cell-intrinsic processes mostly have a constant rate throughout life, i.e., DNA mutations caused by them linearly accumulate with age.

In **experimental models of tumorigenesis** often tumor-promoting chemicals, such as dimethylbenzanthacene or 12-O-tetradecanoyl-phorbol-13-acetate, are applied to the skin of lab animals. The multiple steps of tumorigenesis, from benign papillomas up to malignant carcinomas, can be observed. Repeated treatment of the same skin spots accelerates the process to only a few months. The tumor-promoting compounds either directly cause mutations to genomic DNA or they initiate signal transduction pathways that activate oncogenes, such as *KRAS*. Well-known cancer risk factors, such as smoking and UV radiation, similarly activate pathways related to tumor promotion. Non-repaired mutations to genomic DNA that spread in a malignant tumor cell population are irreversible, while the activation of signal transduction pathways can be reversed by the tumor-promoting compound. This indicates that severe forms of cancer are preventable by avoiding the exposure to tumor-promoting carcinogens (Sect. 1.5). Similarly, not only chemicals can act as carcinogens but also lifestyle factors, such as high fat diet leading to obesity associated with chronic inflammation, can act as tumor promoters. These lifestyle factors may either stimulate the endogenous mutation process or act epigenetically by changing the chromatin landscape (Chap. 6).

Most malignant tumors develop over a period of 20 to 30 years, since the sequential accumulation of two to six driver mutations in the same cell population requires time (Fig. 4.3, bottom). These mutations provide the cancer cells with selective growth advantages and related cellular properties, which are summarized as hallmarks of cancer (Sect. 2.4). In addition to driver mutations, which enhance the activity of oncogenes and reduce that of tumor suppressor genes, the cells accumulate hundreds to thousands of passenger mutations that have no effect on the process of neoplasia. **Passenger mutations occur randomly over time in every tissue and every person,** i.e., **also in the 50% of us who never get the diagnosis of cancer**.

Cancer is recognized as a disease with a very individual profile. Accordingly, malignant tumors even from the same type of tissue can be very distinct concerning their genetic alterations. However, in most types of cancer only a limited number of cell fate controlling pathways are affected by the divergent mutations in driver genes (Sect. 2.2). The hallmarks of cancer concept provides a general explanation of the process of tumorigenesis by postulating that a primary malignant tumor has to acquire all these cellular properties, in order to get fully malignant, i.e., before it starts to evade to distant tissues and forms metastases (Sect. 2.4). However, **the order in which these hallmarks are acquired is flexible**, i.e., there are multiple ways to obtain a malignant tumor. The combination of both observations can explain the molecular and cellular basis of personalized tumorigenesis (Fig. 4.4). This model allows any chronological order, in which the hallmarks of cancer are acquired. However, there is the tendency that the hallmarks "self-sufficiency in growth signals" and "insensitivity to anti-growth signals" are in most cases at the beginning of the tumorigenesis process, while the hallmark "tissue invasion and metastasis" often occurs at the end. Nevertheless, the model implies the option that a primary malignant tumor forms metastases that do not have accumulated all hallmarks of cancer. Therefore, rather small primary malignant tumors may be able to disseminate their cells, some of them even before they had been diagnosed. However, these aggressive metastatic cells may be more sensitive to targeted therapy (Sect. 11.2).

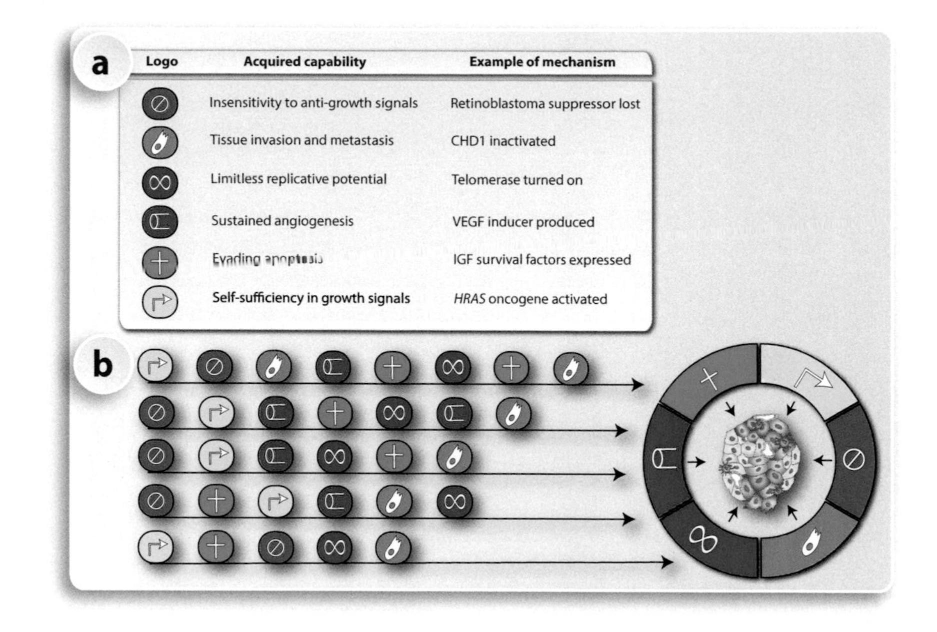

Fig. 4.4 **Personalized tumorigenesis.** In order to explain the very individual history of every case of cancer, it is assumed that basically all permutations of the chronologically order are allowed, in which the six core hallmarks of cancer are acquired. This implies that the two to six driver mutations occur randomly out of hundreds to thousands of options, how a cell may be transformed. Nevertheless, only a limited number of cell fate controlling signal transduction pathways are affected by these genetic alterations. Thus, **cancers are very divergent in their genetic basis and the chronology of these events, but they are convergent concerning the affected pathways and acquired hallmarks**. CDH1 = cadherin 1, also called E-cadherin, IGF = insulin-like growth factor

4.3 Genome Instability

Every day there are about 70,000 mutations to the genomic DNA of every cell in our body (Fig. 4.5). Some 75% of these mutations are single-strand DNA breaks that are based on oxidative damage during metabolism or hydrolysis of bases. In contrast, there are only some 25 double-strand breaks per day and cell, but these bear a higher risk of destroying information within the genome. The remaining 25% of the mutations are some 12,000 events of depurination, some 2800 8-oxo modifications to guanines, approximately 600 cases of depyrimidination and 192 cytosine deaminations. These mutations are often transversions of C (cytosine) into T (thymine) and A (adenine) into G (guanine), which are due to spontaneous deamination of C into U (uracil) and by misincorporation by DNA polymerases during replication and repair. Each of the daily damages has the potential to be converted into a permanent mutation, in case it does not get repaired. However, based on long-lived stem cells and neurons the yearly mutation rate is only approximately 40, i.e., **less than 1 in 600,000 initial mutations are not corrected**.

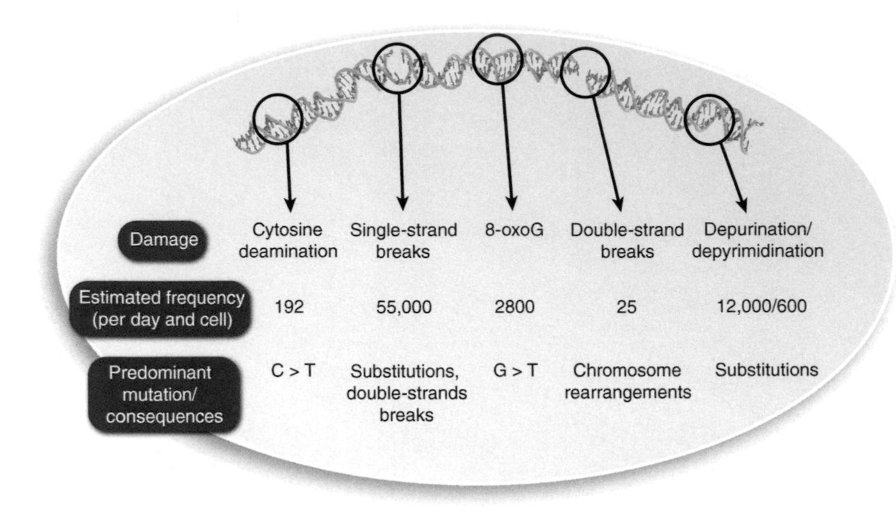

Fig. 4.5 Estimated frequencies of DNA lesions and mutations associated with dysfunctional DNA repair. Details are provided in the text

Fortunately, healthy cells have a number of very effective pathways, in order to repair the large burden of daily DNA lesions:

- **base excision repair (BER)**: correction of modified bases
- **mismatch repair (MMR)**: detection and removal of bases that are misincorporated during DNA replication
- **nucleotide excision repair (NER)**: removal of bulky adducts in DNA
- **homologous recombination (HR)**: double-strand break repair via the use of homologous information of the homologous siter chromosome
- **non-homologous end-joining (NHEJ)**: repair of double-strand breaks without the use of a DNA template.

In BER, MMR and NER damaged DNA strips are excised and new bases are inserted, in order to fill the gap, i.e., these pathways need the activity of a DNA polymerase. NHEJ has a rather high error rate, since it frequently results in insertions or deletions at the breakpoint. In contrast, double-strand repair via HR has a far higher fidelity, since it is controlled by the non-mutated allele. All DNA repair pathways are mediated by key proteins that are referred to as "caretakers", because they either detect DNA damage and activate the repair machinery, repair damaged DNA or inactivate mutagenic molecules before they can harm the genome. Accordingly, many **genes encoding for caretaker proteins are tumor suppressors**, such as *TP53*, *BRCA1* and *BRCA2*.

The loss of DNA repair genes leads to an increase in the mutation rate over the whole genome. This enhanced **genome instability** is an emerging hallmark of cancer and contributes to tumorigenesis. Hereditary forms of cancer (Sect. 5.1) are often based on mutations in caretaker genes involved in DNA repair pathways. For example, hereditary breast and ovarian cancers are usually due to mutations in the genes *BRCA1*, *BRCA2* and *PALB2* (partner and localizer of BRCA2), which control the double-strand break repair by HR. Moreover, advanced prostate cancer often comprises driver mutations in HR-mediated repair genes, such *BRCA2* and *ATM*, and MMR genes, such as *MLH1* (MutL homolog 1) and *MLH2*. The observed tissue specificity is not understood, since essential DNA repair genes, such as *BRCA1* and *BRCA2*, are ubiquitously expressed. In contrast, the specificity of mutations in genes of the NER pathway, such as *ERCC5* (ERCC excision repair 3, TFIIH core complex helicase subunit), causing the skin cancer predisposition syndrome xeroderma pigmentosum is explainable by the fact that only skin is exposed to UV radiation.

The loss of telomeric DNA can cause genome instability, such as amplification and deletion of chromosomal segments, so that the telomerase complex (Sect. 7.5) can be considered as a caretaker. However, *TERT* is an oncogene, since its activation can lead to immortality of cancer cells. In a normal cell all DNA repair pathways provide a feedback to caretakers, such as p53 (Sect. 3.1), and arrest their cell cycle or stimulate apoptosis in case of non-reparable DNA lesions. In contrast, cancer cells tolerate the complexity of their genome caused by loss-of-function of caretaker genes, such as *TP53*. Accordingly, **genome instability is rather the result of cancer than its cause**.

The classification of DNA mutations (Box 4.3) based on type of SNV, type of rearrangement, such as duplication, inversion, deletion or translocation, and their sequence context distinguishes a spectrum of mutation categories and their relation to deficiencies in one or the other DNA repair pathway. For example, mutations are more evenly distributed over the genome, when either the NER or the MMR pathway is deficient. Moreover, defects in HR, MMR and NER pathways have signatures that are associated with specific types of cancer, such as a higher mutation rate in genomic regions located within accessible chromatin in melanoma, ovarian, lung and prostate cancer. For example, such mutational signatures can distinguish whether:

- a lung cancer came from a smoker or nonsmoker
- a hepatocellular carcinoma derived from exposure to aflatoxin
- a melanoma was caused by UV radiation
- ovarian cancer is based on *BRCA1* or *BRCA2* mutations.

Interestingly, mutational signatures can direct therapeutic decision making. For example, a deficiency in the MMR pathway massively increases the occurrence of neoantigens, which significantly increases the success of immunotherapies (Sect. 10.2).

Box 4.3: Types of DNA mutations. Functionally most important are SNVs within the coding sequence of proteins. **Synonymous** mutations do not alter the encoded protein, while **non-synonymous** mutations cause a change in the amino acid sequence (**missense**) or introduce a pre-mature stop codon (**nonsense**). **Indels** as well as **copy number variations** (CNVs) in exonic sequences can result in either non-frameshift or frameshift mutations. Moreover, CNVs in intronic sequences may lead to alternative splicing. Other genetic variations in the non-coding region of the genome affect transcription factor binding sites

Taken together, **defects in genome maintenance and repair provide another selective advantage for tumor progression**, since they speed up the rate at which pre-malignant cells can accumulate more favorable genotypes.

4.4 Cancer Driver Mutations and Genes

All cells of our body accumulate mutations over time, i.e., the older we get the more our genome deviates from its sequence at birth. This accumulation derives from the slightly imperfect repair of the huge number of daily DNA lesions and over lifetime accounts in every cell of our body for a few thousand mutations (Sect. 4.3). However, our tissues differ in their exposure to environmental carcinogens. Therefore, the mutation rate is higher in skin, colon and lung than in brain or heart (Fig. 4.6).

Only a few percent of the mutations found within a somatic cell affect protein coding sequences and a subset of these are non-synonymous (Box 4.3), i.e., they change the amino acid sequence. These mutations have the highest potential to be **driver mutations**. Accordingly, an average adult solid cancer, such as from colon, breast, pancreas or brain, carries some 50 genes with mutations that alter their encoded protein products (Fig. 4.6, center). However, in melanoma or lung cancer more than 200 genes with non-synonymous mutations are found reflecting their exposure to potent mutagens, such as UV radiation and tobacco smoke (Fig. 4.6, left). Importantly, lung cancers from smokers have 10-times more non-synonymous mutations than that of non-smokers. Moreover, cancers with mutations in DNA repair pathways (Sect. 4.3) have a far higher mutation rate than other cancers, i.e., also endogenous factors can affect the number of mutations. In contrast, pediatric cancers have in average less than 10 genes with non-synonymous mutations (Fig. 4.6, right).

The vast majority of somatic mutations are considered as **passengers**, since they do not change the functionality of the concerned genes, so that they enhance the growth potential of cells. In contrast, driver mutations affect driver genes in a way that they provide a cell clone with at least one additional hallmark of cancer resulting in a selective advantage in comparison with its neighboring cells. Cancer driver genes are classified as oncogenes (Chap. 2), when at least 20% of their mutations lead to gene activation, and as tumor suppressor genes (Chap. 3), if more than 20% of genomic alterations cause gene inactivation. The 20% rule is important, since not every mutation of a cancer driver gene has a major effect and for some genes the effects of their mutations may point into opposite directions. When this rule was applied to more than

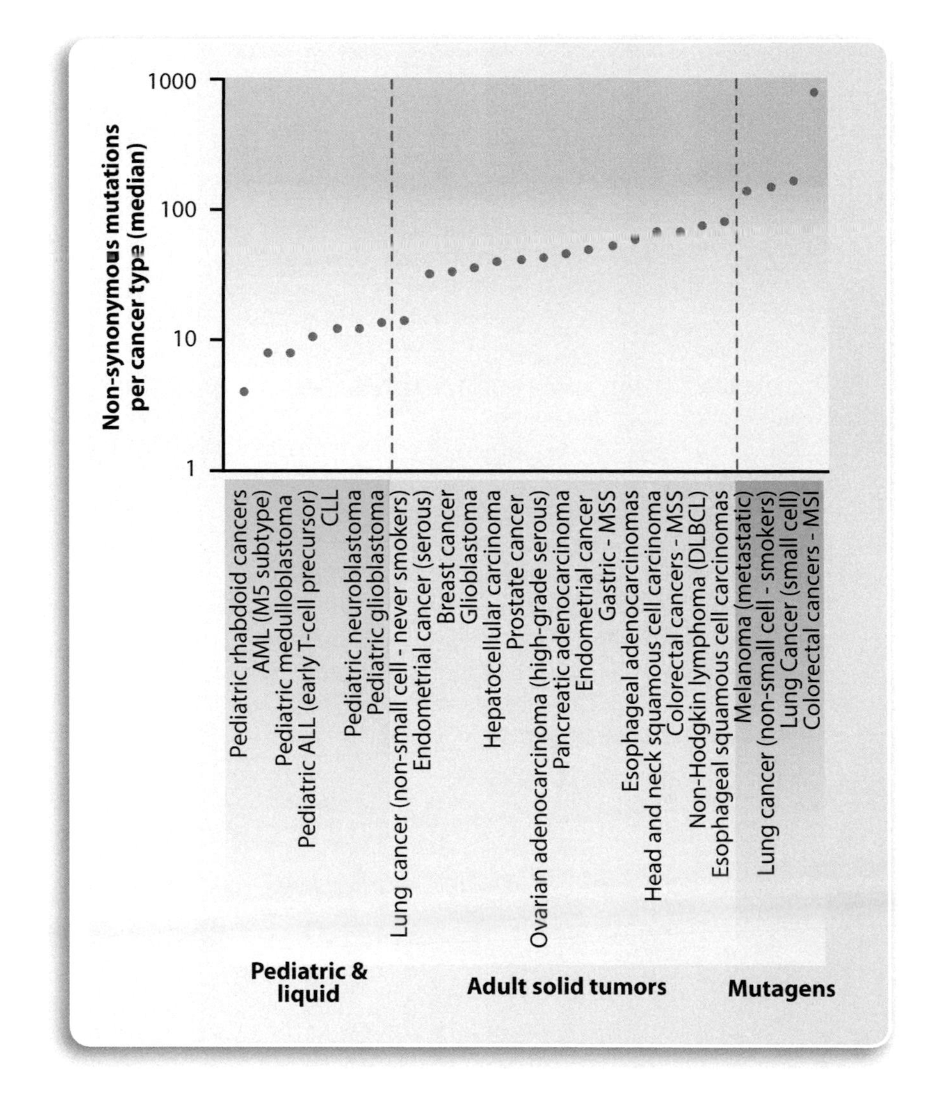

Fig. 4.6 Somatic mutations in human cancers. The median number of non-synonymous mutations per cancer were compared for pediatric and liquid cancers (**left**) and adult cancers (**right**). Data are based on Science *339*, 1546-1558. DLBCL = diffuse large B cell lymphoma, MS = myeloid sarcoma, MSI = microsatellite instability, MSS = microsatellite stable

3000 human malignant tumor samples, 125 cancer driver genes were identified, 54 of which are oncogenes and 71 tumor suppressor genes. There are also a few unclear cases. For example, *NOTCH1* acts in lymphomas and leukemias as an oncogene and in squamous cell carcinomas as tumor suppressor gene. The main mutations to the initial list of 125 cancer driver genes are SNVs, i.e., base substitutions and indels. However, there are also additional cancer driver genes, which are not point-mutated but changed by genomic rearrangements, such as translocations, amplifications and larger deletions. Based on whole exome sequencing data from 11,873 pairs of cancer and normal controls provided by cancer genome projects (Sect. 5.3) 460 driver genes had been identified that clustered into 21 cancer-related pathways. However, **the huge complexity of many cancer genomes, which is primarily based on large chromosomal rearrangements and duplications, often makes the identification of the key driver genes of a given malignant tumor very difficult**.

Between tissues there is a large variation for their risk of developing cancer (Sect. 1.4). Some of these differences are related to the environmental exposure with carcinogens or infections, which may account in total for some 25% of the overall cancer risk. In addition, for some 5–10% of all cancers a strong inherited genetic predisposition is the main driver (Sect. 5.1). However, **the main contribution to the risk for cancer seems to be stochastics**, i.e., in many tissues the chance of getting cancer is convincingly correlated to the number of stem cell divisions (Fig. 4.7). Since the risk of accumulating mutations is proportional to the number of DNA replication events that result in non-corrected errors, a tissue that constantly renews and therefore shows a higher number of stem cell divisions, should be more vulnerable to cancer. This implies that, e.g., colorectal cancer risk would be far lower, if colon cells would not constantly divide. Moreover, stochastics also determines, whether the inevitable somatic mutations that we accumulate in our tissues affect only passenger genes or also driver genes. Thus, **some 2/3 of our lifetime cancer risk seems to be good or bad luck**. The consequence of this insight is that for many cancer cases there may be no primary prevention possible but major emphasis should be taken on secondary

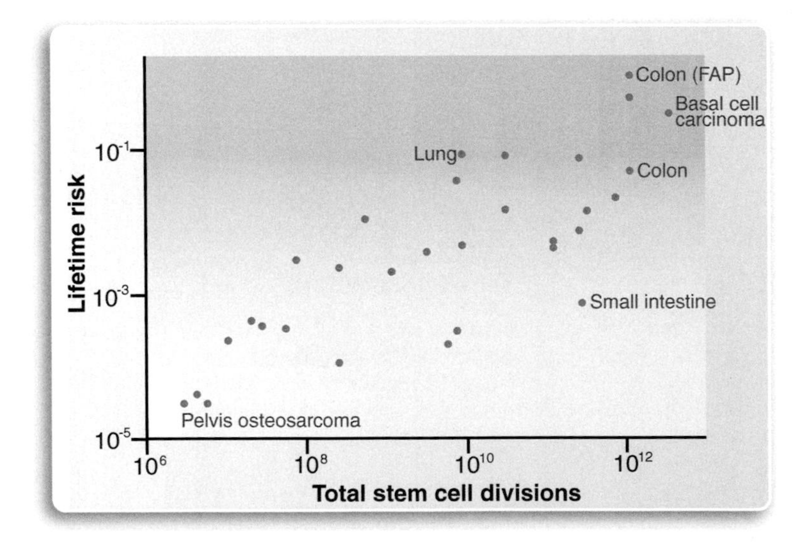

Fig. 4.7 Stem cell division and cancer risk. The graph indicates for 32 cancer types a correlation between the number of stem cell divisions and the risk of cancer. Data are based on Science *347*, 78-81

prevention, such as early detection (Sect. 1.5). This also implies the efficiency of the immune system to recognize and eliminate cancer cells (Sect. 10.1). Fortunately, in an increasing number of cases the diagnosis of cancer does not mean that one will die from the disease (Sect. 1.1). This may apply to a larger extent to cancer types that largely occur by chance than for those that are caused by environmental factors. Thus, even if only 25% of all cancer cases are primarily driven by lifestyle decisions, they may represent 50% of all deadly cancers. Accordingly, **primary cancer prevention is still very important**.

Clinical conclusion: Tumorigenesis is a multi-step process behind every type of cancer. It is based on the accumulation of cancer driver mutations in the same cell(s) and normally takes years to decades. Thus, it needs to be obvious to everyone that a healthy lifestyle should be a basic attitude for cancer prevention. This includes participation in cancer screening programs.

Further Reading

Balmain, A. (2020). The critical roles of somatic mutations and environmental tumor-promoting agents in cancer risk. *Nature Genetics, 52,* 1139–1143.

Kumar, S., Warrell, J., Li, S., McGillivray, P. D., Meyerson, W., Salichos, L., Harmanci, A., Martinez-Fundichely, A., Chan, C. W. Y., Nielsen, M.M., et al. (2020). Passenger mutations in more than 2,500 cancer genomes: overall molecular functional impact and consequences. *Cell, 180,* 915–927 e916.

McGranahan, N., & Swanton, C. (2017). Clonal heterogeneity and tumor evolution: past, present, and the future. *Cell, 168,* 613–628.

Tomasetti, C., & Vogelstein, B. (2015). Cancer etiology. Variation in cancer risk among tissues can be explained by the number of stem cell divisions. *Science, 347,* 78–81.

Tubbs, A., & Nussenzweig, A. (2017). Endogenous DNA damage as a source of genomic instability in cancer. *Cell, 168,* 644–656.

Chapter 5
Cancer Genomics

Abstract Cancer is a disease of our genome, since it is primarily caused by sporadic mutations of the DNA in our cells. In addition, like in most common diseases, we all carry inherited variations of our genome that determine our susceptibility to different types of cancer. The development of next-generation sequencing methods allowed so far the systematic sequencing of the genomes of primary malignant tumors from up to 38 different cancer types. At present, cancer genome projects have identified 576 cancer driver genes on the basis of some 30,000 malignant tumors. Continuing these efforts to millions of sequenced cancer genomes will significantly improve our understanding of the tumorigenesis process and will allow more efficient precision oncology.

Keywords SNPs · Cancer susceptibility · Genetic architecture · Next-generation sequencing · Genomics · Driver genes · Catastrophic mutational events

5.1 Human Genetic Variation and Cancer Susceptibility

Whole genome sequencing has indicated that **every individual carries, on average, 4 million genetic variants** (Box 5.1) covering about 12 million base pairs (Mb) of sequence (0.3% of all). Most of these genetic variants are neutral, i.e., they do not contribute to phenotypic differences or disease risk, and achieved simply by chance significant frequencies within respective human populations. In this context, the sum of rare, high-penetrance variants is of significant influence. However, there is a large number of common variants with a small to modest effect size that have a dominant role in common complex traits. Hundreds of complex phenotypic traits determine how an individual looks and behaves as well as his/her risk to develop non-communicable diseases including the different types of cancer. Furthermore, each complex trait is based on dozens to hundreds of gene variants and environmental influences, i.e., for the understanding of the molecular basis of a trait its **genetic architecture** needs to be uncovered, such as:

- the number of variants that influence a heritable phenotype
- their relative magnitude concerning different traits
- the population frequency of the respective variants

C. Carlberg and E. Velleuer, *Cancer Biology: How Science Works*,
https://doi.org/10.1007/978-3-030-75699-4_5

- the interactions of the variants with each other and the environment.

Box 5.1: Human genetic variation. The different types of human genetic variants are referred to as common (or polymorphisms), when they have a minor allele frequency (MAF) of at least 1% within the studied population, or as rare, when they have a MAF of less than 1%. **SNPs** represent the most common class of genetic variations among individuals and approximately 7 million SNPs show a MAF of more than 5% (www.ncbi.nlm.nih.gov/SNP). The *1000 Genomes Project* indicated that humans have in total more than 88 million genomic variants, i.e., in addition there is a huge number of rare and novel SNPs. Follow-up projects increased this number to nearly 500 million. Nevertheless, the majority of variants of any given individual are common in the whole population. In parallel, some 60,000 unique CNVs are known and some of them are quite common in human populations. Since the detection of structural variants needs advanced technology, basically all initial associations between genome variations and complex traits, such as observed by GWAS, were done only with SNPs. Nevertheless, per individual structural variants cover between 9 and 25 Mb of sequence, i.e., 0.3-0.8% of the whole genome. In average, **a typical human genome contains some 150 SNVs resulting in protein truncation, 10,000 SNVs changing amino acids and 500,000 SNVs affecting transcription factor binding sites**. Interestingly, each individual is heterozygous for 50-100 genetic variants that can cause inherited disorders in homozygous offspring, including different types of cancer or cancer predisposition syndromes. This provides a large demand and challenge for genetic counseling based on **whole genome sequencing**. Moreover, gene-environment interactions provided by lifestyle factors, such as the personal choice of food, will create an additional level of complexity

Mendelian disorders are monogenetic, for which, in most cases, a single SNP can explain the occurrence of the disease. However, common diseases, including cancer, have a multigenic basis and were investigated by **genome-wide association studies (GWASs)**. These studies employ an agnostic approach in the search for unknown disease variants, i.e., hundreds of thousands of SNPs are tested for association with a disease in large cohorts of patients *versus* healthy controls. Disease-associated genomic loci can be found in the database GWAS Catalog (www.ebi.ac.uk/gwas). GWASs with 2000 to 5000 individuals confidently identified common variants with effect sizes, referred to as odds ratios, of 1.5 or greater, i.e., a 50% increased risk for the tested disease. Larger sample sizes were achieved by pooling several GWASs through meta-analyses. For example, sample sizes of at least 60,000 subjects provide sufficient power to identify the majority of variants with odds ratios of 1.1, i.e., a 10% increased risk. However, despite some notable successes in revealing numerous novel SNPs and genomic loci associated with complex phenotypes, **most of polygenic traits have less than 20% of their heritability explained by the common variants**.

The impact of SNPs on the protein coding sequence of our genome (some 1.5% of all) is well established. Non-synonymous mutations cause a change in the amino acid sequence (missense) or introduce a pre-mature stop codon (nonsense). Indels as well as larger copy number variations in exonic sequences can result in frame-shift mutations. Moreover, variations in intronic sequences may lead to alternative splicing. However, the vast majority of genetic variants are located in regulatory and not in coding regions of genes, i.e., **the phenotypic consequences of most genetic variants are rather based on an epigenetic or gene regulatory processes than on a change in protein function**.

Common SNPs of a complex multigenic disease, such as cancer, are characterized by low odds ratios (Fig. 5.1, right), while rare monogenetic forms of the disease have high odds ratios (Fig. 5.1, left). Examples of the latter are mutations in the tumor suppressor genes:

- *BRCA1* and *BRCA2* associated with hereditary breast and ovarian cancer
- *TP53* as the basis of the Li-Fraumeni syndrome leading to cancers of the breast, brain and adrenal glands
- *RB1* producing familial retinoblastoma
- *APC* in the FAP syndrome leading to colorectal cancer.

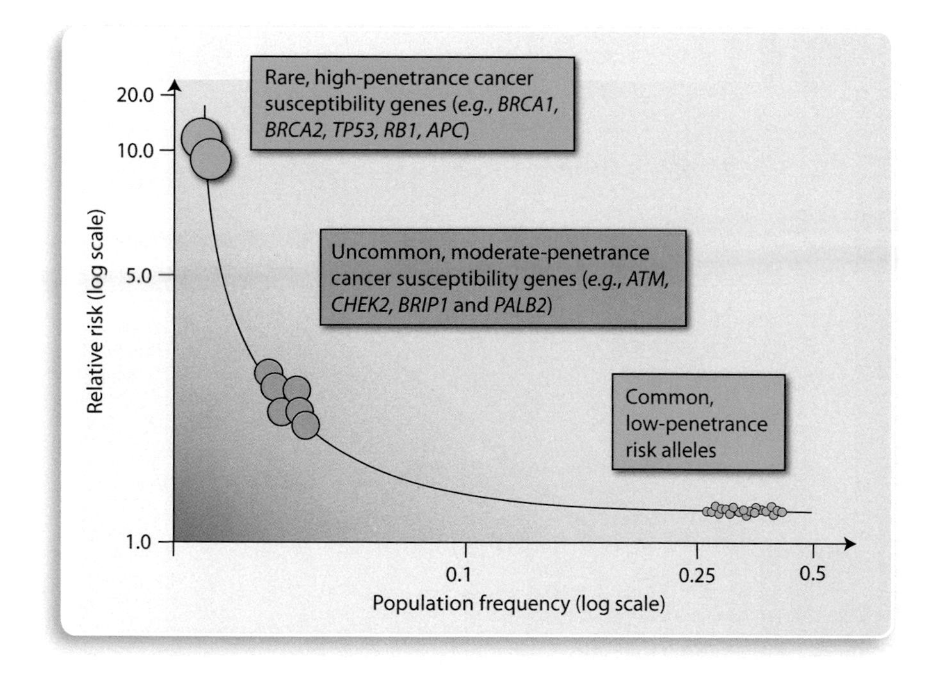

Fig. 5.1 Genetic architecture of cancer risk. The graph displays the relation of odds ratios for cancer risk (y-axis) in relation to frequency of a SNP (x-axis) as identified by GWAS. Examples for high- and mid-penetrance genetic variants are indicated

It is expected that whole genome sequencing of large numbers of individuals will identify a higher count of low frequency SNPs with intermediate odds ratios (Fig. 5.1, center), which may better explain an individual's genetic disease risk. Known examples of the latter are SNPs in the genes *ATM* (the basis of the cancer predisposition disease Ataxia-telangiectasia), *CHEK2* (link to Li-Fraumeni syndrome), *BRIP1* (BRCA1 interacting protein C-terminal helicase 1, also called *FANCJ,* link to the cancer predisposition syndrome Fanconi anemia) and *PALB2* (also called *FANCN*, link to Fanconi anemia).

In industrialized countries with the culture of high-fat nutrition (Western diet) the lifetime risk of a female to get breast cancer is 1 in 8, i.e., the average breast cancer susceptibility for this population of more than 500 million females is 13%. From these, 5–10% carry a strong hereditary risk for developing the disease (Fig. 5.2, left), of which the tumor suppressor genes *BRCA1* and *BRCA2* have with approximately 20% the largest known contribution. Other mid-penetrance risk genes and their relative contribution are not yet well characterized. Thus, the vast majority of breast cancer cases appear to occur sporadically (Fig. 5.2, right). However, also in these cases there is an interaction of low-penetrance risk genes with environmental factors, such as lifestyle decisions on diet and exercise. Similar scenarios apply to basically all types of cancer, i.e., also other **cancers have an average hereditary rate of 5–10%**.

Every person's genome can be analyzed far before the possible onset of cancer (Sect. 5.2), i.e., knowing one's individual cancer risk may have an important value

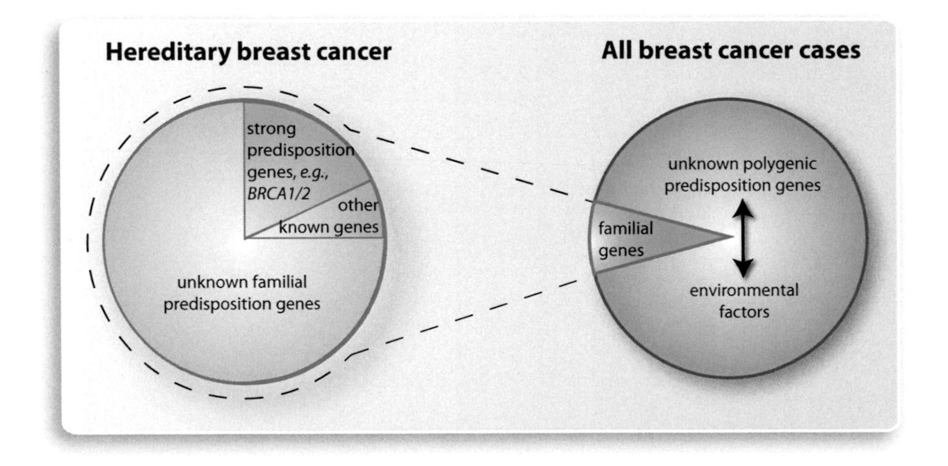

Fig. 5.2 Breast cancer susceptibility. Hereditary breast cancer (**left**) represents only about 5-10% of all breast cancer (**right**). The tumor suppressor genes *BRCA1* and *BRCA2* account for about 20% of the familial risk, i.e., for less than 2% of the total risk. Most of the genetic variants that contribute to the risk of developing sporadic breast cancer are unknown but they interact with environmental factors, such as life(style) decisions on diet, exercise and number of births combined with breast feeding

for disease prevention, such as triggering intensive screening programs or prophylactic surgery, such as removing breast tissue and ovaries of females with inherited *BRCA1* or *BRCA2* mutants. At present there are more than 100 high-penetrance cancer predisposition genes known that have either a cancer type-specific profile, such as *VHL* mutations driving renal cancers, or a broader range (Fig. 5.3).

5.2 The Cancer Genome

The clinical application of next-generation sequencing methods (Box 5.2) initially involved targeted sequencing of some 500 selected cancer genes (cancer gene panel) or of all protein coding genes (the exome, comprising only 1.5% of the genome). Nowadays, decreasing costs make sequencing of the whole genome (3200 Mb) of a cancer clone affordable along with a healthy control from the same patient. This allows the identification of mutations, such as point mutations, indels, CNVs and translocations, in all regions of the genome. In this way, not only the key driver mutations are identified but also the complete set of passenger mutations. Most of the driver mutations are non-synonymously affecting the protein coding sequence of oncogenes and tumor suppressor genes (Sect. 4.4). Knowing the set of driver genes of a given cancer sample will allow targeted therapy, in case a specific drug, such as a kinase inhibitor or a neutralizing antibody, has already been developed (Sect. 11.2). Moreover, whole genome sequence information is in particular important for the identification of epimutations (Chap. 6).

Box 5.2: Next-generation sequencing. The development of massively parallel DNA sequencing methods, which are also referred to as next-generation sequencing, allows the simultaneous sequencing and analysis of millions of short DNA fragments in parallel. Technological advances in high-throughput sequencing significantly decreased the costs within the past 15 years to below 1000 Euros per whole genome. This makes the sequencing of a malignant tumor sample alongside with normal tissue, mostly obtained from white blood cells, of the same patient affordable. In comparison with reference genomes, both somatic mutations obtained during the lifetime of the patient as well as germline variations are detected. Information obtained by genomic analysis of tens of thousands cancer patients has not only helped in disease stratification and in the identification of their molecular mechanisms (Fig. 5.3), but also transforms the perspective of future health care from disease diagnosis and treatment to personalized health monitoring and preventive medicine (Sect. 11.3)

Sequencing healthy and cancer genome can be of impact in any phase of life (Fig. 5.4). The information obtained from a whole genome sequence can be important

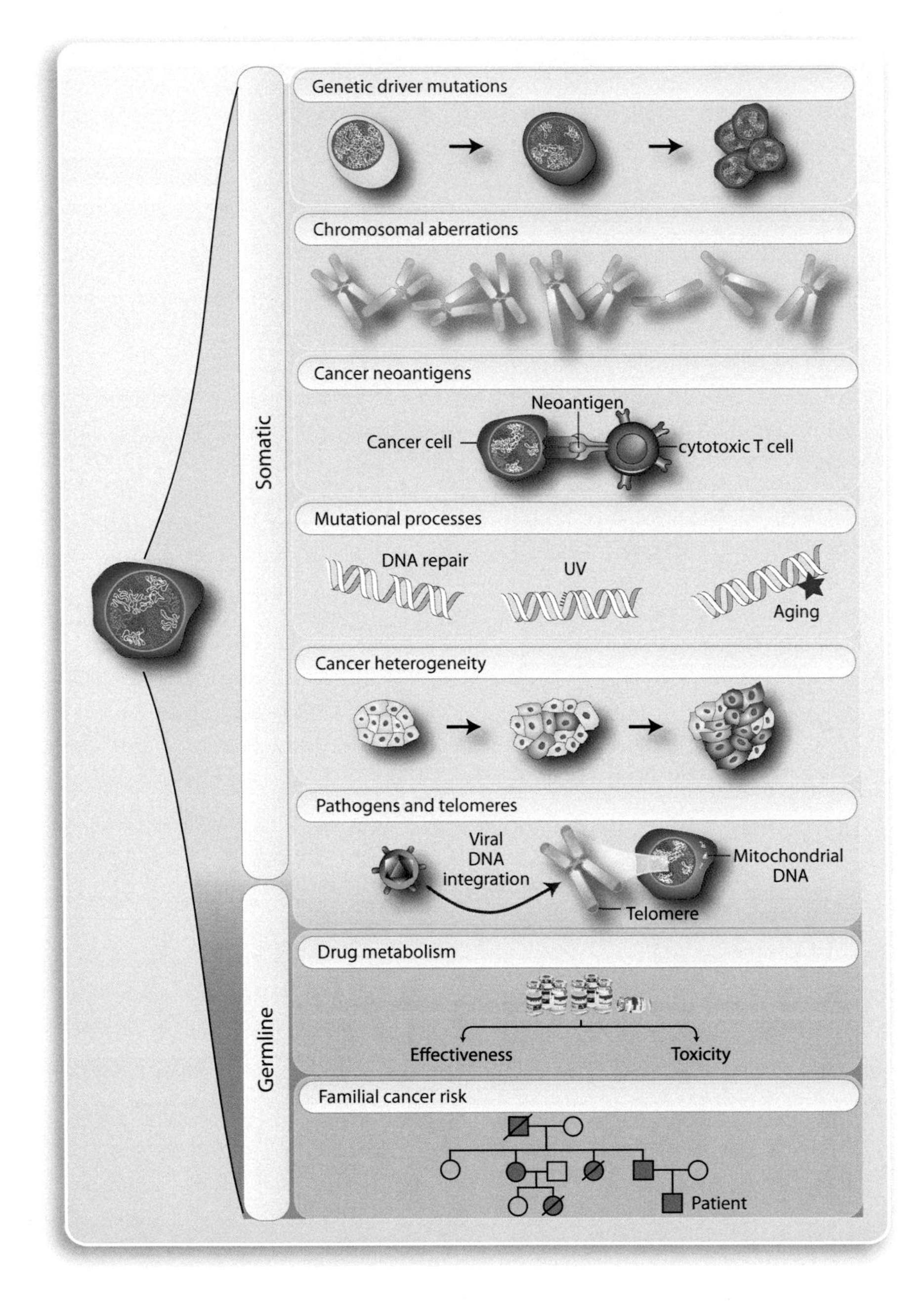

Fig. 5.3 Insights from cancer genome analysis. Whole genome sequencing of cancer clones allows to understand many aspects of cancer biology in molecular detail. These are the identification of driver mutations, large-scale chromosomal abnormalities and neoantigens, i.e., mutations that may be recognized by the immune system (Sect. 10.2). Moreover, in comparison with reference genomes inherited variations in cancer genes are detected as well as in metabolic enzymes that may influences the efficacy of therapeutic compounds

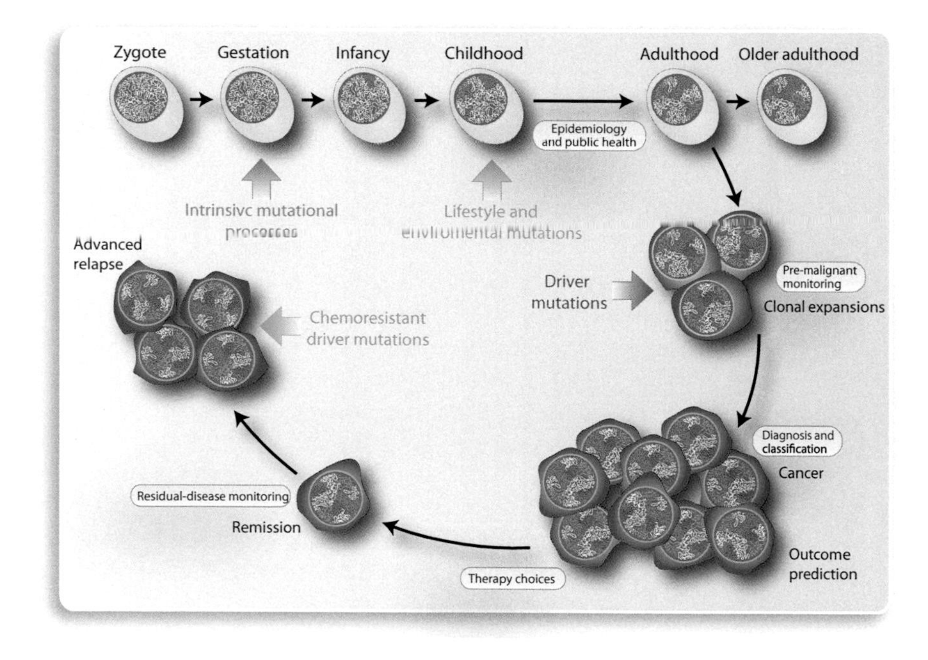

Fig. 5.4 Impact of cancer genome sequencing. The information obtained from sequencing of the genome of a cancer can influence the cancer management strategies on multiple levels. More details are provided in the text

for:

- public health cancer prevention initiatives
- early interventions before a cancer becomes invasive
- strategies for the diagnosis, classification, treatment and monitoring of established cancers.

Mutational signatures of different cancer types based on cancer genome sequences will allow a better understanding of the molecular basis for different incidence rates based on geographic region as well as on occupational and lifestyle exposures. The sequencing of early cancer lesions, e.g., in prostate cancer, will allow a stratification on the basis of the number of driver mutations and CNVs into those that need a treatment and others that may grow very slow or may even regress spontaneously ("watch and wait"). **This could avoid an overtreatment**. Complete sequence information on a cancer will support its histological classification (Box 4.2) and should lead to a more detailed categorization. A genomic classification of a cancer has the clear advantage that it is based on the disease causing driver mutation. This may suggest a precise targeted treatment, in case an appropriate drug is available (Sect. 11.3). Moreover, detailed genomic information will allow a more accurate prognosis of different cancer types, which is very important for deciding on therapeutic options, some of which may be very risky due to toxic side effects. In this context, monitoring possible changes of the cancer clone via genome sequencing can get of high impact, in order to prevent treatment resistance due to overgrowth by a clone with new driver mutations or genomic rearrangements.

5.3 Cancer Genome Projects

Molecular biologists followed the example of physicists and realized that some of their research aims could only be reached through multi-national collaborations of dozens to hundreds of research teams and institutions in so-called **big biology projects** (Box 5.3). The Cancer Genome Atlas (TCGA) was the first big biology project in the cancer field and had characterized within 12 years more than 10,000 cases of primary cancers and their respective normal controls representing 33 types of cancer (Fig. 5.5). This huge amount of data was even extended by combing them with that of other projects under the umbrella of the International Cancer Genome Consortium (ICGC, https://daco.icgc.org). A common goal of these projects was the systematic and comprehensive identification of the key genes driving malignant tumors via whole exome and gene sequencing identifying SNVs and CNVs. Moreover, TCGA systematically collected transcriptome,

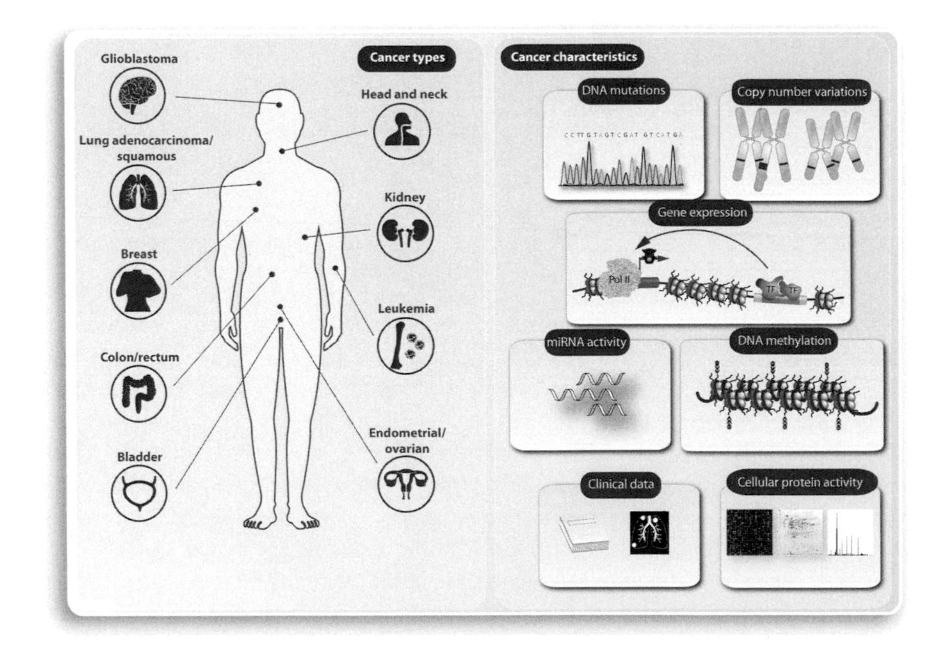

Fig. 5.5 The TCGA project. The TCGA research network had initially chosen the indicated 12 major types of cancer (later extended to 33) representing thousands of patients and systematically investigated the indicated characteristics of their cancers

epigenome, miRNA expression, protein activity and clinical data of the patients and their malignant tumors.

Box 5.3: Big biology projects. The *Human Genome Project* (www.genome.gov/human-genome-project), which was launched in 1990 and finished in 2003, was the first example of a big biology project. Together with follow up studies the project had a tremendous impact on the understanding of the architecture and function of human genes. Improved next-generation sequencing methods (Box 5.2) allowed personal genome sequencing of both normal and cancer genomes. This made large-scale genome sequencing studies possible, such as the *1000 Genomes Project* (www.1000genomes.org) and *TCGA* (www.cancer.gov/about-nci/organization/ccg/research/structural-genomics/tcga). Furthermore, the *Encyclopedia of DNA elements (ENCODE) Project* (www.genome.gov/encode) and focused on the functional characterization of the human genome. The *ENCODE* follow-up project *Roadmap Epigenomics* (www.roadmapepigenomics.org) provided human epigenome references from 111 primary human tissues and cell lines

With decreasing sequencing cost TCGA, ICGC and other cancer projects shifted from whole exome to whole genome analysis, 2658 samples of which are investigated in context of the PCAWG (PanCancer Analysis of Whole Genomes) consortium (https://dcc.icgc.org/pcawg). These samples represented 38 malignant tumor types and their integrated analysis identified 44 million SNVs, thousands of which were identified as cancer drivers with an average of 4.6 driver mutations per malignant tumor (87% within in protein coding sequence). Interestingly, in 9% of the cases no known driver gene mutations were identified, i.e., **cancer driver discovery is not yet complete**. At present (February 2021), the Cancer Gene Census (https://cancer.sanger.ac.uk/census) lists 576 cancer driver genes, 536 of which show somatic mutations, 102 germline mutations, 156 frameshift mutations, 253 missense mutations, 155 nonsense mutations, 73 splicing mutations, 314 translocations, 24 amplifications and 43 large deletions. Most of these genes function only in one or two cancer types as drivers, while a few, such as *TP53*, *PIK3CA*, *KRAS*, *PTEN* and *RB1*, are cancer-wide drivers. Some cancer driver genes are found at high frequency only in one or two cancer types, such as mutations in *MYC* and *CCND3* in 60 and 47% of all Burkitt lymphomas (Sect. 2.3), respectively. Interestingly, cancer types differ in their preference for point mutations and structural variations. For example, breast adenocarcinomas show in average 6.4 driver CNVs and only 2.2 SNVs, whereas colorectal adenocarcinomas have only 2.4 CNVs and 7.4 SNVs.

Cancer genome projects confirmed that *TP53* is by far the most mutated cancer gene occurring in about half of all investigated primary malignant tumors. For 77% of all *TP53* driver mutations they supported the two hit concept for tumor suppressor inactivation or deletion (Sect. 3.3). Together with *CDKN2A* and *MYC*, *TP53* is one of the few cancer driver genes that is ubiquitously expressed. Interestingly, in some cancer types *TP53* mutations are often found together with *KRAS* mutations or *MYC* amplifications, i.e., **the concepts of cancer cell transformation of the pre-genomic era were endorsed** (Chaps. 2 and 3). The very detailed cataloging of mutations in *TP53* and other cancer driver genes provided links between typical mutations and their source, such as the R249S amino acid exchange in p53 protein in hepatocellular carcinoma due to aflatoxin exposure or the R213STOP mutation in melanoma based on UV radiation. Genome-wide the DBD of p53 is the protein domain being affected most often by driver gene mutations followed by tyrosine kinase domains of 13 different kinases, such as BRAF (Sect. 11.2). Narrow clusters of mutations are found for codons 12 and 13 of KRAS in 85% of colorectal cancer cases or codon 132 of isocitrate dehydrogenase 1 (IDH1) in 100% of AML cases. Interestingly, non-coding mutations of driver genes are much less frequent than those of the protein coding region. However, the regulatory region of the *TERT* gene, which often gets significantly upregulated in malignant tumors obtaining the hallmark "limitless replicative potential", is a prominent exception.

Cancer genome projects also allowed the investigation of complex chromosomal rearrangements, so-called catastrophic mutational events (Box 5.4), in malignant tumors. The process of **chromoplexy** was found in 17.8% of all samples, in particular in prostate and thyroid cancers as well as in lymphoid malignancies. Deletions and complex rearrangements were strongly associated with **kataegis**, which was detected even in 60.5% of all cancers showing highest rates in lung squamous cell carcinoma, bladder cancer, acral melanoma and sarcomas. Finally, **chromothripsis** was observed in 22.3% of all investigated malignant tumors, most frequently in melanoma. The latter process of genomic rearrangement was often associated with *TP53* inactivation or *TERT* overexpression. Interestingly, chromothripsis often occurs early in tumorigenesis, especially in liposarcomas, prostate adenocarcinoma and lung squamous cell cancer.

Box 5.4: Catastrophic mutational events. Often complex chromosomal rearrangements of the genome reflect the accumulation of DNA damage over time, but they can also arise from "all at once" catastrophic events. The three major types are

- **chromoplexy**: a mutational process where genomic regions from one or more chromosomes become scrambled, i.e., are broken and ligated to each other in a new configuration
- **kataegis**: a pattern of localized hypermutations, in which a large number of highly patterned SNVs occur in a small genomic region of only a few hundred bp. It results from the activity of the family of APOBEC (apolipoprotein B mRNA editing catalytic subunit) enzymes that act as mutators via the deamination of C generating U
- **chromothripsis**: a mutational process by which up to thousands of clustered chromosomal rearrangements occur in a single event found in specific

genomic regions of one or multiple chromosomes. It often occurs in cancer cells with a dysfunctional p53 pathway

Taken together, during the past 15 years cancer genome projects studied some 30,000 tumor samples and have increased the number of known driver genes from 10–20 to 576 (Fig. 5.6). This may represent more than 90% of all driver gene. It is expected that cancer genomics is shifting into a consolidation phase, where the sequencing of millions of further malignant tumor genomes will lead only to a minor increase in the number of driver genes. So far, cancer genomics studied primary malignant tumors and drivers therein, while it is possible that future investigations of metastatic tumors will uncover new genes and mechanisms. For example, the genes estrogen receptor (*ESR1*) and androgen receptor (*AR*) are rarely mutated in primary breast and prostate tumors, respectively, but there are well-known mutational drivers of tumors with resistance to anti-estrogens and anti-androgens. At present,

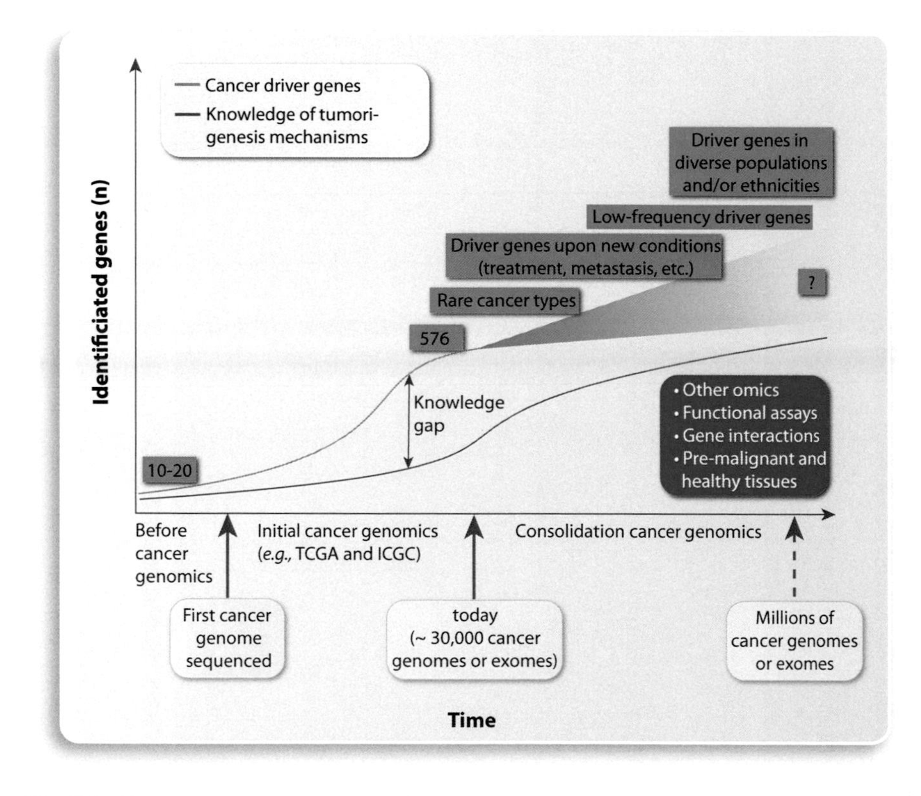

Fig. 5.6 Perspective on cancer genomics. The identification of the first 10-20 cancer drivers genes dates before the start of the cancer genomics era leading after some 15 years to 576 genes. It is expected, that cancer genomics is now transferring to a consolidation phase, in which the knowledge gap to the mechanisms of tumorigenesis is closing

there is still a significant knowledge gap between identifying cancer driver genes and comprehensively understanding their function in the tumorigenesis process. Therefore, a major challenge of cancer biology projects is to fill this gap in the future. Thus, **insight into tumorigenesis is essential for translating the knowledge on cancer genomics into efficient precision oncology** (Sect. 11.3).

Clinical conclusion: Genome-wide sequence information on cancer driver gene mutations are essential for a precise treatment of cancer and will revolutionize cancer diagnosis.

Further Reading

ICGC, TCGA consortium. (2020). Pan-cancer analysis of whole genomes. *Nature, 578,* 82–93.

Martínez-Jiménez, F., Muiños, F., Sentís, I., Deu-Pons, J., Reyes-Salazar, I., Arnedo-Pac, C., et al. (2020). A compendium of mutational cancer driver genes. *Nature Reviews Cancer, 20,* 555–572.

Nangalia, J., & Campbell, P. J. (2019). Genome sequencing during a patient's journey through cancer. *The New England Journal of Medicine, 381,* 2145–2156.

Chapter 6
Cancer Epigenomics

Abstract Compared to normal cells, cancer cells show epigenetic drifts, which are genome-wide changes in DNA methylation, histone modifications and 3-dimensional (3D) chromatin structure representing epimutations. Moreover, many malignant tumors reactivate programs of fetal development indicating that tumorigenesis is associated with epigenetic reprogramming. The mechanistic bases of cancer epigenomics are specific genetic, environmental and metabolic stimuli that disrupt the homeostatic balance of chromatin, which then either becomes very restrictive or permissive. Very most types of cancers display a clear epigenetic contribution to the different hallmarks of cancer suggesting that epigenetic modulators, modifiers and mediators form an additional classification system for cancer genes.

Keywords Chromatin · Chromatin modifying enzymes · Epimutation · Epigenetic reprogramming · DNA methylation · Histone modifications · Chromatin architecture · Cell state transitions · Epigenetic mediators · Epigenetic modulators

6.1 Epigenetic Mechanisms of Cancer

Genomic DNA is wrapped around complexes of histone proteins that help to fit the genome into a cell nucleus with a diameter of less than 10 μm. This protein-DNA complex is referred to as **chromatin** and composed of regularly repeating **nucleosomes** (Fig. 6.1, center). Chromatin is organized into lower-order structures, such as the 11 nm fiber of **euchromatin** and higher-order structures, like the 30 nm fiber of **heterochromatin** or the 700 nm fibers of chromosomes occurring only during the metaphase of mitosis. A cell's phenotype depends on its gene expression pattern, which basically is influenced how the genomic DNA is packed into chromatin. The regularly positioned nucleosomes (one every 200 bp) often block the access of transcription factors to their genomic binding loci, since the packing of genomic DNA around them hides one side of the DNA. In order to get transcription factor binding sites accessible and their genes activated, **chromatin remodeling enzymes**, such as SWI/SNF (switching/sucrose non-fermenting), are used to either induce only minor shifts in the position of the nucleosomes or to deplete in other cases a whole nucleosome (Fig. 6.1, bottom).

C. Carlberg and E. Velleuer, *Cancer Biology: How Science Works*,
https://doi.org/10.1007/978-3-030-75699-4_6

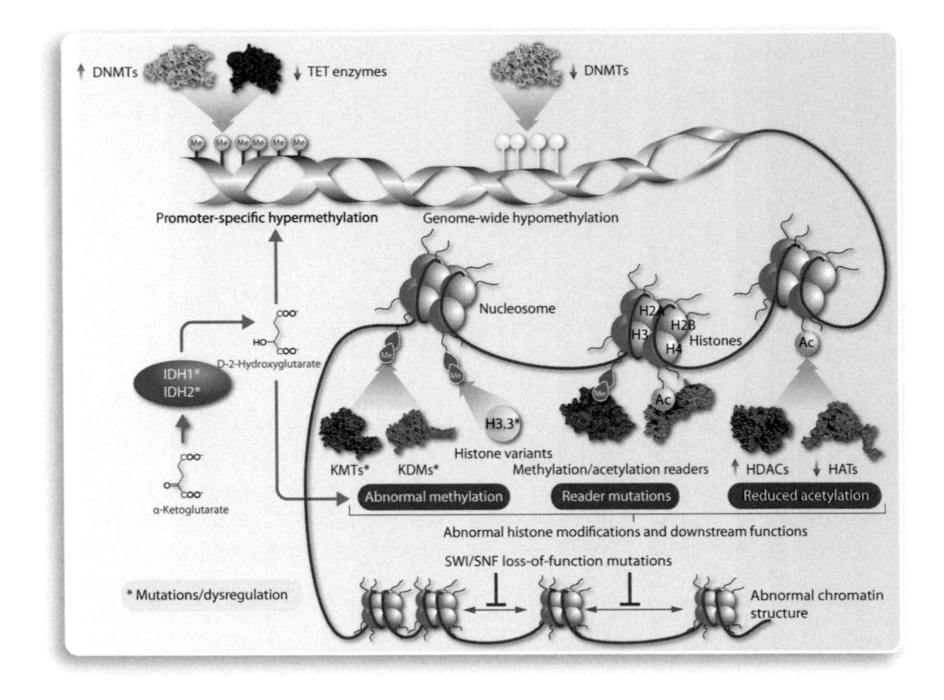

Fig. 6.1 Epimutations in cancer. There are four major types of epimutations affecting cancer: DNA hypermethylation at promoters (**top left**), genome-wide DNA hypomethylation (**top right**), abnormal modification of histones and/or their recognition (**center**) and abnormal chromatin structures caused by mal-functional chromatin remodelers (**bottom**). More details are provided in the text

The tails and globular domains of the nucleosome-forming **histone proteins** H2A, H2B, H3 and H4 as well as the linker histone H1 provide over 130 amino acid residues for post-translational modifications, such as phosphorylation, acetylation and methylation, and the introduction of histone variants (Box 6.1, Fig. 6.1, center). Post-translational modifications of histones are frequent and important epigenetic signals that control many biological processes, such as cellular differentiation in the context of embryogenesis. The content of these information-processing circuits is summarized as the **histone code**. These genomic locus-specific histone modifications are reversible and an important component of the epigenetic memory affecting transcription factor binding and differential gene expression between cell types. Thus, **nucleosomes are not only barriers that block access to genomic DNA but serve as dynamic platforms linking and integrating many biological processes**, such as transcription and replication.

Box 6.1: Chromatin organization. Chromatin acetylation is generally associated with transcriptional activation, whereas the exact residue of the histone tails that is acetylated is not very critical. In contrast, histone methylation, as mediated by KMTs, mainly mediates chromatin repression, but at certain residues, such as H3K4, it results in activation. Therefore, for histone methylation, in contrast to acetylation, the exact position of the residue within the histone tail as well as its degree of methylation (mono-, di- or tri-methylation) is of critical importance. The effects of chromatin modifiers, such as HATs and HDACs, are primarily local and may cover only a few nucleosomes up and downstream of the starting point of their action. The same applies to KMTs, KDMs and chromatin remodeling enzymes, such as the SWI/SNF complex. When there is more HAT activity, chromatin is locally acetylated, the attraction between nucleosomes and genomic DNA decreases and the latter gets accessible for activating transcription factors, basal transcription factors and Pol II. In this euchromatin state chromatin remodeling enzymes may have to fine-tune the position of the nucleosomes in order to obtain full accessibility of the respective binding sites. In the opposite case, when HDACs are more active, acetyl groups get removed and the packing of chromatin locally increases. KMTs then methylate the same or adjacent amino acid residues in the histone tails that attract heterochromatin proteins, such as HP1 (heterochromatin protein 1), and further stabilize the local heterochromatin state. The structure and organization of chromatin can be interpreted as a number of superimposed epigenetic layers that lead either to open **euchromatin** and active gene expression or to closed **heterochromatin** and no gene expression:

- the DNA methylation status at CG dinucleotides (CpGs), where hypermethylation stimulates the formation of heterochromatin (Sect. 6.2)
- the packing of nucleosomes, where more dense arrangements indicate heterochromatin (Sect. 6.3)
- histone modifications at specific positions mark for either active chromatin (mainly acetylated) or inactive chromatin (mainly methylated)
- the accessibility of genomic DNA for the binding of transcription factors
- the complex formation and relative position of the chromatin, such as active transcription factories in the center of the nucleus and inactive chromatin attached to the nucleoskeleton at the nuclear periphery (Sect. 6.3)

Within an average terminally differentiated human cell only some 100,000 chromatin loci are open while more than 90% of the genome is buried in heterochromatin and not accessible to transcription factors and polymerases. However, **accessibility of these chromatin regions is not static as they are dynamically controlled by chromatin modifying and remodeling proteins**. These enzymes catalyze the methylation of genomic DNA, the post-translational modification of histone proteins or the positioning of nucleosomes. Our genome expresses in a tissue-specific fashion hundreds of these chromatin modifiers and remodelers that recognize (read), add (write) and remove (erase) chromatin marks. Writer-type enzymes, such as **histone acetyltransferases** (HATs), **lysine methyltransferases** (KMTs) and **DNA**

methyltransferases (DNMTs), add acetyl or methyl groups to histone proteins or cytosines of genomic DNA, respectively. In contrast, eraser-type enzymes, such as **histone deacetylases** (HDACs), **lysine demethylases** (KDMs) and **DNA demethylases** (TET (ten-eleven translocation) enzymes), reverse these reactions and eliminate the respective marks (Fig. 6.1, top and center). Reader-type proteins, such as methyl-binding domain (MBD) proteins and the chromatin organizing protein CTCF (CCCTC binding factor), bind genomic DNA depending on its methylation status. Moreover, also components of the chromatin remodeling complexes are able to read chromatin marks.

Dysregulated epigenetic processes can act as drivers of early disruption of cellular homeostasis in pre-cancerous and cancerous cells in the tumorigenesis process (Sect. 4.2). An important epigenetic change in cancer is the deregulation of CpG methylation patterns, i.e., of the DNA methylome (Sect. 6.2). Like in aging (Sect. 7.2), tumorigenesis correlates with genome-wide DNA hypomethylation (Fig. 6.1, top right), which leads to genome instability via the reactivation of pluripotency transcription factors (acting as oncogenes, Sect. 6.4) and retrotransposons within repetitive genomic DNA (Box 1.1). In contrast, CpG islands at promoter regions of tumor suppressor genes get hypermethylated (Fig. 6.1, top left, Sect. 6.2), which causes the inactivation of the respective genes and their tumor-protective function. This is representative for an **epigenetic drift**. For example, silencing of the tumor suppressor gene *MLH1* via DNA hypermethylation harms the MMR DNA repair process and increases the risk of accumulating DNA mutations throughout the whole genome. Thus, **one epimutation can initiate a multitude of genetic changes**.

Mutations in genes encoding for chromatin modifiers are either gain-of-function or loss-of-function. Abnormal histone methylation may be caused by mutations in genes encoding for KMTs and KDMs as well as for the histone variant H3.3, which reduces the genome-wide methylation of H3K27 and H3K36 (Fig. 6.1, center). Examples are gain-of-function and overexpression of EZH2 (enhancer of zeste 2 polycomb repressive complex 2 subunit, a H3K27-specific KMT) and loss-of-function of the H3K36-specific KMT SETD2 (SET domain containing 2). Moreover, there are translocations of KMT2A (H3K4-specific) as well as translocations and overexpression of the H3K36-specific KMT NSD1 (nuclear receptor binding SET domain protein 1) and the H3K27-specific KMT NSD2. In addition, the amplification or overexpression of genes encoding for H3K4-, H3K9- and H3K36-specific KDMs have been described in the context of different types of cancer. Furthermore, also histone acetylation is reduced in cancer through the loss of the HATs EP300 (E1A binding protein p300, also called KAT3B) and CREBBP (CREB binding protein, also called KAT3A) and the overexpression of HDACs. Finally, **not only the writer and eraser function of chromatin modifiers can be affected by mutations but also their reader function**. Examples are the overexpression or gain-of-function translocations of BRD4 (bromodomain containing 4), which binds acetylated histones, or the overexpression of TRIM (tripartite motif containing) 24, which recognizes H3K23ac. In many cancers, histone proteins or their variants, such as H3.3, are mutated (Fig. 6.1, center). For example, 90% of chondroblastomas and 20% of pediatric glioblastomas have a

K36M mutation in histone H3.3, which inhibits H3K36-specific KMTs, reduces H3K36 methylation and alters gene expression.

Loss-of-function mutations in genes encoding for DNA demethylases (*TET1*, *TET2* and *TET3*) or increased expression of genes encoding for DNMTs (*DNMT1*, *DNMT3A* and *DNMT3B*) can cause promoter hypermethylation in some cancers (Fig. 6.1, top). In contrast, genome-wide hypomethylation is often based on loss-of-function mutations in the *DNMT3A* gene. Mutations in the genes encoding for the metabolic enzymes IDH1 and IDH2 cause that the citric acid cycle intermediate α-ketoglutarate becomes transformed into the oncometabolite 2-hydroxyglutarate, which inhibits TETs and KDMs (Fig. 6.1, center left). This leads to the increased methylation of both DNA and histones. Taken together, **epimutations affect a large variety of changes in cellular homeostasis that result in the acceleration of tumorigenesis**.

The cancer genome and epigenome influence each other in a multitude of ways and can work mutually. Both genetics and epigenetics offer complementary mechanisms to achieve similar results. For example, a gain-of-function activation of the oncogene *PDGFRA* (platelet-derived growth factor receptor α) that is important for achieving the hallmark "sustained proliferative signaling" may be based either on a genetic mutation within the coding region of the gene or by an epimutation that disrupts the 3D chromatin strip carrying the gene. Moreover, the inactivation of tumor suppressor genes may occur either by a deletion within the coding region or epigenetic silencing of its promoter region (Sect. 3.3). **Epigenome profiling** (Box 6.2) of healthy and diseased cells, such as cancer cells, leads to maps of DNA methylation, histone marks, DNA accessibility and DNA looping that can be visualized with an appropriate web browser, such as the UCSC Genome Browser (https://genome.ucsc.edu).

Box 6.2: Epigenome profiling *via* next-generation sequencing technologies. There are a number of next-generation sequencing (Box 5.2) methods that investigate various aspects of chromatin biology, such as DNA methylation, histone modification state and 3D chromatin structure. Global epigenomic profiling allows hypothesis-free exploration of new observations and correlations. Individual research teams as well as large consortia, such as *ENCODE* and *Roadmap Epigenomics*, have already produced thousands of epigenome maps from hundreds of human tissues and cell types. The integration of these data, e.g., transcription factor binding and characteristic histone modifications, allows the prediction of enhancer and promoter regions as well as monitoring their activity and many additional functional aspects of the epigenome. The key epigenetic methods determine DNA methylation, transcription factor binding and histone modification, accessible chromatin, and 3D chromatin architecture. The biochemical cores of these methods are:

- different chemical susceptibility of nucleotides, such as bisulfite treatment of genomic DNA, in order to distinguish between cytosine and its methylated forms by the use of methylated DNA immunoprecipitation sequencing, or base-resolution mapping methods, such as bisulfite sequencing

- affinity of specific antibodies for chromatin-associated proteins, such as transcription factors, modified histones and chromatin modifiers, as determined by the method chromatin immunoprecipitation (ChIP) followed by sequencing (ChIP-seq)
- endonuclease-susceptibility of genomic DNA within open chromatin compared to inert closed chromatin, such as measured by DNase I hypersensitivity followed by sequencing (DNase-seq) and more recently by assay for transposase-accessible chromatin using sequencing (ATAC-seq)
- for assessing the 3D organization of chromatin proximity ligation of genomic DNA fragments that via looping got into close physical contact, such as measured in chromosome conformation capture (3C)-based methods and its genome-wide version, high-throughput chromosome capture (Hi-C)

Large-scale cancer genome projects, such as *TCGA* (Sect. 5.3), revealed that nearly all human cancers carry mutations in key chromatin-associated proteins. It is important to realize that the epigenetic signature of a cell allows more variation than its primary genetic status. The error rate in inheritance of DNA methylation is some 4% for a given CpG dinucleotide per cell division, while the mutation rate of the genome during DNA replication is far lower (Sect. 4.3). Thus, **epigenetic variability leads in much shorter time to an epimutation and phenotypic selection, such as possible onset of cancer, than genetic mutations in a traditional view**. Interestingly, a child is too young to acquire severe genetic mutations leading to cancer but epigenetic mutations alter much faster the cellular homeostasis. Accordingly, the majority of mutations found in cancers arising in adults are traditional genetic mutations, while many childhood cancers are based on epimutations (Sect. 6.4).

6.2 DNA Methylation and Cancer

The identity of each of our 400 tissues and cell types is based on their respective unique gene expression patterns, which in turn are determined by differences in their epigenomes. For the proper function of our tissues, it is essential that cells memorize their respective epigenetic status and pass it to daughter cells when they are proliferating. **The main mechanism for this long-term epigenetic memory is the methylation of genomic DNA at the 5th position of cytosine (5mC)**. The DNA methylome is a genome-wide map of 5mC patterns and its oxidized modifications (Box 6.3) and represents an essential component of the epigenome. Compared with normal cells of the same individual, the DNA methylome of cancer cells shows a massive overall loss of DNA methylation, while for certain genes also hypermethylation at CpGs is observed. This so-called CpG island methylator phenotype (CIMP) is

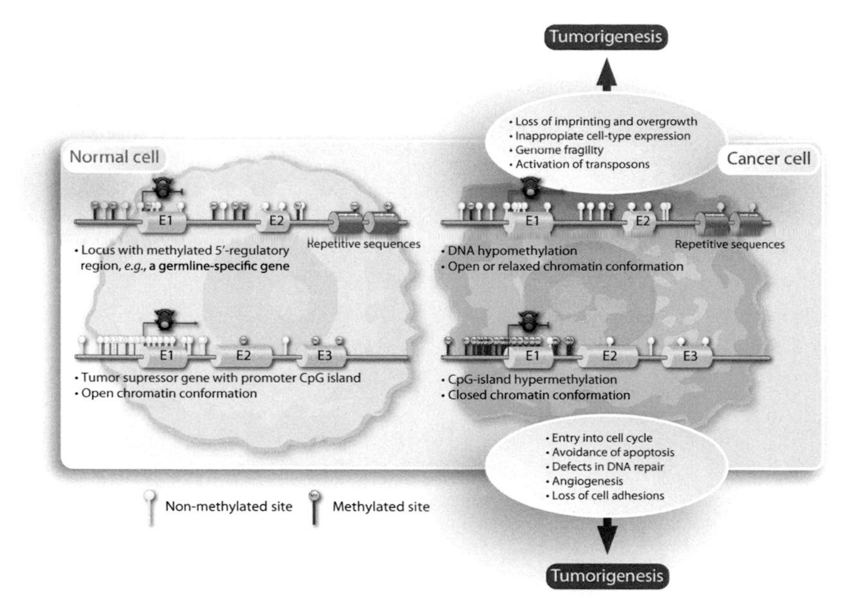

Fig. 6.2 **Changes of DNA methylation patterns during tumorigenesis.** Compared to normal cells (**left**), cancer cells are hypomethylated on the genome-wide scale (**top right**), in particular at repetitive sequences, such as transposons. In addition, imprinted and tissue-specific genes often get demethylated. Hypomethylation causes changes in the epigenetic landscape, such as the loss of imprinting, and increases the genome instability that characterizes cancer cells. Another common alteration in cancer cells is the hypermethylation of CpGs within regulatory regions of tumor suppressor genes (**bottom right**). These genes are then transcriptionally silenced so that cancer cells lack functions, such as inhibition of the cell cycle

the best-known epigenetic dysregulation in cancer. Thus, **changes in DNA methylation patterns are key epigenetic dysregulations occurring during tumorigenesis** (Fig. 6.2).

Box 6.3: The human DNA methylome. The average CG percentage of our genome is 42%, i.e., one strand of the haploid genome contains some 700 million cytosines. In principle, each of them can be methylated, but CpG dinucleotides are symmetrically methylated on both the top and bottom DNA strand. Therefore, only the methylation pattern of CpGs persists during DNA replication and can be inherited to both daughter cells. Since **the DNA methylome of somatic cells represents an epigenetic program of global repression of the genome and specific settings of imprinted genes, it is important to maintain it during replication. CpG islands** are defined as genomic regions of at least 200 bp in length showing a CG percentage higher than 55%, i.e., only a minority (10%) of all CpG dinucleotides belong to CpG islands. Many of the approximately 28,000 CpG islands within the human genome are found rather close or within promoter regions of protein-coding genes. In contrast, the majority of CpG dinucleotides are located within regions of **repetitive genomic DNA**, such as SINEs, LINEs and LTRs (Box 1.1). LINEs and LTRs carry strong promoters that must be constitutively silenced *via* placing them into constitutive heterochromatin. Methylated genomic DNA is transcriptionally repressed, i.e., in most cases there is an inverse correlation between DNA methylation of **regulatory genomic regions**, such as promoters and enhancers, and the expression of the genes that they are controlling (Fig. 6.2). However, at their **gene bodies** highly expressed genes show high levels of DNA methylation, i.e., some methylated CpGs downstream of TSS (transcription start site) regions positively correlate with gene expression. Genes driven by CpG-rich promoters are silenced when methylated, while genes without CpG islands close to their TSS regions are regulated by other mechanisms than DNA methylation, such as transcription factors binding to enhancers. The silencing of the repetitive DNA happens primarily during early embryogenesis, while in adult tissues *de novo* silencing is initiated by **methyl-binding proteins**, such as MECP2 (methyl-CpG-binding protein (2), MBD1, MBD3 and MBD4. MBD proteins are not classical transcription factors but act as readers and adaptors for the recruitment of chromatin modifiers, such as HDACs and KMTs, to methylated genomic DNA

DNMTs are chromatin modifying enzymes that catalyze in a one-step reaction the transfer of a methyl group from S-adenosyl-L-methionine (SAM) to cytosines of genomic DNA (Fig. 6.1, top). The main responsibility of DNMT1 is the **maintenance of the DNA methylome** during replication. During early embryogenesis, i.e., in the pre-implantation phase, most CpGs are unmethylated. After implantation DNMT3A and DNMT3B **de novo methylate** those CpGs that had not been packed with H3K4me3-marked nucleosomes. Approximately 28% of patients with AML carry mutations in the *DNMT3A* gene, some which reduce catalytic activity of the encoded enzyme and lead to focal hypomethylation.

Differential DNA methylation is established by de novo methylation combined with active demethylation of CpG islands. The latter is a multi-step process that involves the methylcytosine dioxygenase enzymes TET1, 2 and 3, which convert 5mC to 5hmC (5-hydroxymethylcytosine), which is deaminated to 5-hydroxyuracil (5hmU). 5hmU:G mismatches are recognized and removed by the enzyme TDG (thymine-DNA glycosylase) and the abasic site is then repaired by the BER pathway (Sect. 4.3) and results in the overall demethylation of the respective cytosine. Thus, **the oxidative modification of 5mC via the TET/TDG pathway allows a dynamic regulation of DNA methylation patterns**. Interestingly, the first approved epigenetic anti-cancer drug, decitabine (5-aza,2′-deoxy-cytidine), is used for the therapy of leukemia and other forms of blood cancer, in which hematopoietic progenitor cells do not maturate. Decitabine acts as an anti-metabolite that blocks DNA methylation via the inhibition of DNMTs (Sect. 11.2).

Global **DNA hypomethylation** during tumorigenesis generates chromosomal instability, reactivates transposons and causes loss of imprinting. The resulting low DNA methylation favors mitotic recombination leading to deletions and promotes chromosomal rearrangements, such as translocations. The disruption of genomic imprinting, such as the loss of imprinting of the *IGF2* gene, is a risk factor for different types of cancer, such as colorectal cancer or Wilms' tumor. Furthermore, **hypermethylated promoter regions** of tumor suppressor genes, such as *TP53*, *RB1* and *MGMT* (O-6-methylguanine-DNA methyltransferase), can serve as biomarkers that provide significant diagnostic potential in the clinic, in particular in early detection screenings of individuals with a high familial risk of developing cancer (Sect. 5.1). Many CpGs can become methylated already early in tumorigenesis, in particular in CIMP of colorectal cancer, glioma and neuroblastoma. The profiles of CpG hypermethylation vary with cancer types. Each type of cancer can be characterized by its specific DNA hypermethylome, i.e., these epigenetic marks are comparable to traditional genetic and cytogenetic markers. From about 200 genes that are regularly mutated in various forms of human breast and colorectal cancers, on average 11 carry a mutation in a single malignant tumor type. For comparison, 100-400 CpGs close to TSS regions are found to be hypermethylated in a given malignant tumor, i.e., **epigenetics is able to provide 10-times more information than genetics**.

Based on the model of an **epigenetic landscape** (Sect. 1.3), the epigenetic status of a cell, such as its methylation level, can be represented by a ball trapped in a valley. In case of normal differentiated cells, the borders of the valley are high and gene regulatory networks keep the cells in stable epigenetic homeostasis (Fig. 6.3a). This prevents the epigenetic state from moving too far from its equilibrium point in normal tissue. In contrast, a dysregulation of the epigenome during tumorigenesis, such as overexpression of an epigenetic modulator or an inflammatory insult, flattens the valley (Fig. 6.3b). Under these conditions of reduced regulation, the epigenetic status is more relaxed and influenced by stochastic variations. Thus, during tumorigenesis, DNA methylation levels diffuse away from the initial state in normal cells (Fig. 6.3c). A plot of CpG methylation levels during the transformation of normal cells into adenoma and carcinoma cells shows a tight distribution in normal tissue, but a progression from adenoma to carcinoma. This explains the substantial level of epigenetic variation for a given cancer type across individuals or between metastatic cells originating from the same primary malignant tumor.

Aberrant DNA methylation is not only a well-established marker of cancer and disturbed genomic imprinting, but it also can lead to general instabilities of the genome through reduced heterochromatin formation on repetitive sequences. One of the most frequent mutations found in human diseases is a C to T transition at methylated CpG islands, i.e., the epigenetic mark cytosine methylation induces a genetic point mutation by reducing the efficiency of DNA repair at these sites. DNA methylome profiles, e.g., of white blood cells that can be obtained from test persons with minimal invasion, may serve as biomarkers for evaluating the individual risk of cancer. Moreover, the DNA methylome indicates the progress of aging showing

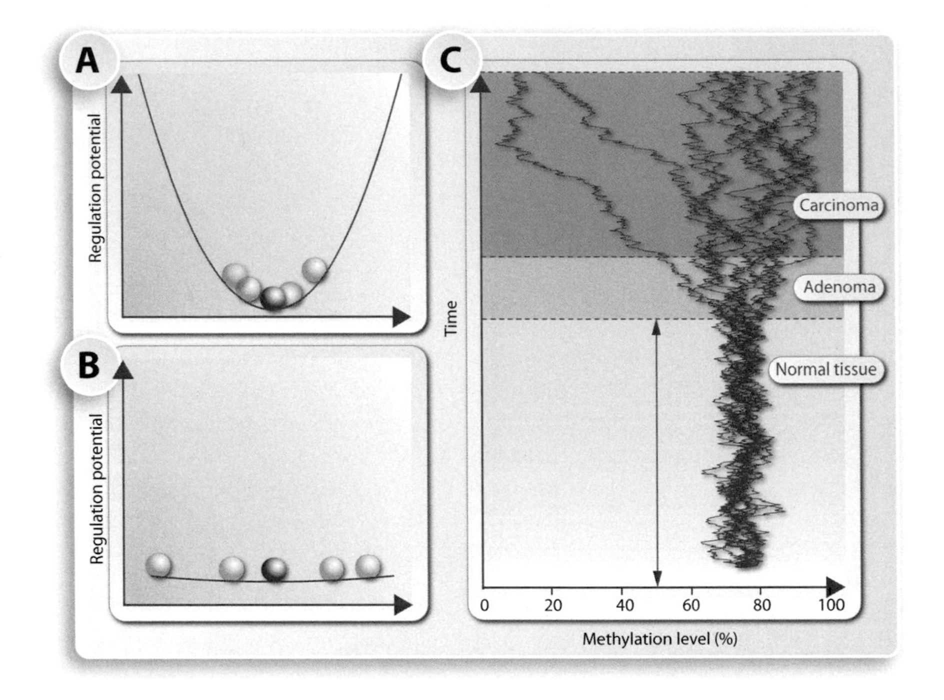

Fig. 6.3 Exemplary model of epigenetic dysregulation. DNA methylation is used here as an example for epigenetic dysregulation. The methylation level of a normal cell is illustrated as a ball at the bottom of a bin (**A**), where regulatory forces, such as gene regulatory networks, allow only minor changes in the epigenetic status. In contrast, during tumorigenesis (**B**) the landscape flattens, and the methylation levels can be far more variable. When the methylation level is modeled over time (bottom to top) for 10 examples (**C**), the variations in the transition from normal tissue to adenoma and carcinoma become obvious as wider methylation ranges

significant interindividual differences (Sect. 7.2). Although biomarkers often do not explain the causality of a disease, they can monitor the disease state and may suggest appropriate therapy. Thus, **epigenomic profiles, such as DNA methylation patterns, in combination with genetic predisposition and environmental exposure, may be prognostic for personal risk of disease onset**.

6.3 Chromatin Changes and Cancer

In general, DNA methylation and histone modifications have different roles in gene silencing. **While most DNA methylation loci represent very stable silencing marks that are seldom reversed, histone modifications mostly lead to labile and reversible transcriptional repression**. For example, genes for pluripotency transcription factors in embryogenesis, such as *POU5F1* (also called *OCT4*) and

NANOG (nanog homeobox) need to be permanently inactivated in later developmental stages, in order to prevent possible tumorigenesis. This happens via H3K9 methylation at unmethylated CpGs on TSS regions of these genes, the attraction of HP1, de novo DNA methylation via DNMT3A and DNMT3B and finally transcriptional silencing for the rest of the life of the individual. In contrast, when in differentiated cells pluripotency genes are silenced only by histone modification, the cells can be rather easily converted to iPS cells.

Changes in DNA methylation during tumorigenesis are always combined with other epigenetic dysregulations, such as aberrant patterns of histone modifications and overall changes in the nuclear architecture, i.e., the overall epigenetic landscape of cancer cells is significantly distorted compared to somatic stem cells or differentiated cells (Fig. 6.4). These alterations in the 3D organization of chromatin exemplify epigenome changes during tumorigenesis. In differentiated cells developmentally repressed genes, i.e., genes that are not needed in a given cell type, are often found within lamina-associated domains (LADs) that constitutively localize close to the nuclear periphery. A significant fraction of these LADs represent so-called large organized chromatin K9-modifications (LOCKs), which are genomic regions that are enriched in repressive H3K9me2 and H3K9me3 histone marks (Fig. 6.4, right). This is further promoted by the recruitment of KDMs and HDACs to the repressive environment of the nuclear envelope and DNA hypermethylation at these regions. Thus, 3D chromatin compaction mediates gene repression during lineage specification and represents a form of **epigenetic memory**. This results in reduced transcriptional noise and provides barriers for dedifferentiation. Although adult stem cells are in a less differentiated state than terminally differentiated cells, also in them specific LAD/LOCK structures are found (Fig. 6.4, left).

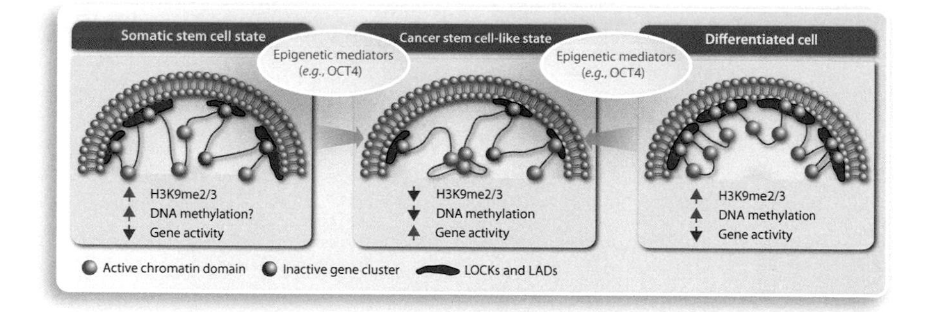

Fig. 6.4 Reprogramming of the nuclear architecture in cancer cells. Epigenetic mediators, such as the transcription factor OCT4, can reprogram the epigenome of somatic stem cells (**left**) or differentiated cells (**right**) into cancer stem cells (**center**). Normal cells are characterized by high levels of H3K9 di- and tri-methylation as well as DNA methylation in LOCKs being a part of LADs. The latter are located close to the nuclear membrane and contain only a low number of active genes. In contrast, in cancer stem cells LOCKs and LADs are largely absent, and a larger variety of genes are active. This leads to phenotypic heterogeneity

Proteins that regulate the interaction of chromatin with the lamina and recruit chromatin modifiers to the nuclear periphery act as epigenetic mediators. For example, the reactivated pluripotency transcription factor OCT4 can reprogram the epigenome of both differentiated cells and adult stem cells into cancer stem cells (Fig. 6.4, center). The activation of the epigenetic mediator dissolves most of the LADs/LOCK structures, as a consequence of which a number of genes are reactivated. This provides cancer cells with phenotypic heterogeneity, such as increased variability in gene expression, in order to switch between different cellular states within the cancer. The loss of LOCKs also affects enhancer-promoter region communication within and between TADs, so that oncogenic super-enhancers are able to cluster. A similar process happens during epithelial-to-mesenchymal transition (EMT), which is a key process in normal wound healing (Sect. 8.1), but also the first step towards metastasis (Sect. 9.1). In EMT the activation of an H3K9-specific KDM, such as KDM1A, often is the initiating epigenetic event. When cancer cells have destabilized the epigenetic memory of the cells they originate from and form EMT-related chromatin structures, they gain phenotypic plasticity. Thus, the overall result of **the change in chromatin architecture might be an oncogenic transformation of the cell**.

The reprogramming of a somatic cell into an iPS cell or its transformation to a cancer cell (Sect. 2.1) are related events. In both cases an epigenetic barrier has to be overcome in a multi-step processes that primarily involves epigenetic mediators, such as the transcription factors SOX2, KLF4, NANOG, OCT4 and MYC or the RNA-binding protein LIN28A (Lin-28 homolog A). Interestingly, all five transcription factors are encoded by oncogenes. Moreover, also the chromatin modifiers SUV39H1 (suppressor of variegation 3-9 homolog 1, also called KMT1A), EHMT2 (KMT1C), SETDB1 (KMT1E), KMT2A, KMT2D, KMT2C, DOT1L (DOT1 like histone lysine methyltransferase, also called KMT4), EZH2 (KMT6A), LSD1 (KDM1A), KDM2B, KDM6A, the SWI/SNF complex member ARID1A (AT-rich interaction domain 1A) as well as DNMTA and DNMT3B have comparable roles both in cellular reprogramming and in tumorigenesis (Fig. 6.5). **Both processes acquire de novo developmental programs and create cells with an unlimited self-renewal potential**.

Changes in cell identity are reflected by alterations in the usage of the enhancer and promoter regions. Many of the regulatory regions that are active in early embryogenesis lose their activity during differentiation. This is compensated through the activity of promoter regions and poised (i.e., paused) enhancers. Changes in enhancer usage require a chromatin topology that allows a new set of enhancers to interact with their target promoters. In parallel, heterochromatin foci become more condensed and more abundant in differentiated cells than in undifferentiated cells. While in ES cells H3K27me3 marks show only focal distributions, in differentiated cells they largely expanded over silent genes and intergenic regions. This results in silencing of pluripotency genes, activating lineage-specific genes and repressing of lineage-inappropriate genes.

Differentiating cells share accessible chromatin regions with the ES cell they derive from, but the similarity in the epigenetic landscape decreases when cells mature. After commitment to a specific lineage, the cellular repertoire expands

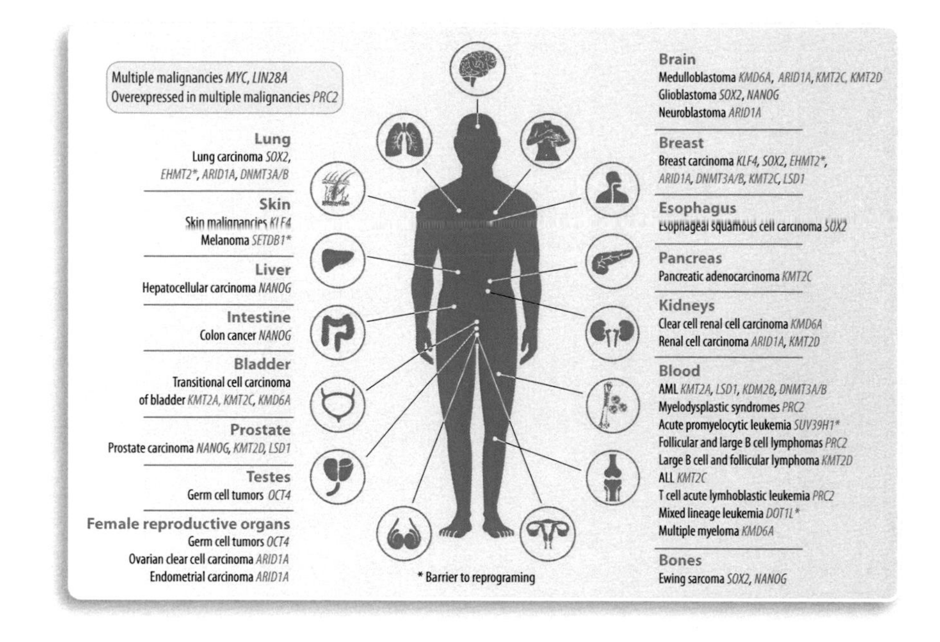

Fig. 6.5 Overlap of iPS nuclear reprogramming and cancer. The same transcription factors (red) and chromatin modifiers (green) play a central role both in iPS cellular reprogramming and in different types of cancer. Some of them are encoded by oncogenes and tumor suppressor genes. SETDB1 = SET domain bifurcated histone lysine methyltransferase 1

for accessible regulatory regions that contain motifs for transcription factors being specific to that lineage, whereas it clearly decreases for transcription factor binding site of other lineages. Thus, the epigenetic landscape of terminally differentiated cells is constrained by the walls of valleys, the height of which are determined by a gene regulatory network. This network is formed by appropriate levels of DNA methylation (Fig. 6.3), histone modifications as well as by a proper 3D architecture (Fig. 6.4). In this way, cells are prevented from switching energy states (Fig. 6.6, top). However, **in response to relevant intra- and extracellular signals, the epigenome also allows cell state transitions**. When chromatin homeostasis is disturbed, e.g., by epimutations (Sect. 6.1), cells do not respond appropriately to these signals. Overly restrictive chromatin networks create epigenetic barriers that prevent all types of cell state transitions (Fig. 6.6, center). In contrast, excessively permissive chromatin networks have very low barriers and allow multiple types of cell state transitions (Fig. 6.6, bottom). For example, deviations from the norm contribute to tumorigenesis.

On the mechanistic level (Fig. 6.6, top) the scenarios of normal, restrictive and permissive chromatin can be explained, e.g., by the actions of a KMT for repressive H3K27me3 marks, such as EZH2, and a KMT for activating H3K4me3 marks, such as KMT2A. EZH2 is the catalytic core of PRC2 (Polycomb repressive complex 2). In normal cells both KMTs and their histone marks are in balance resulting in bivalent,

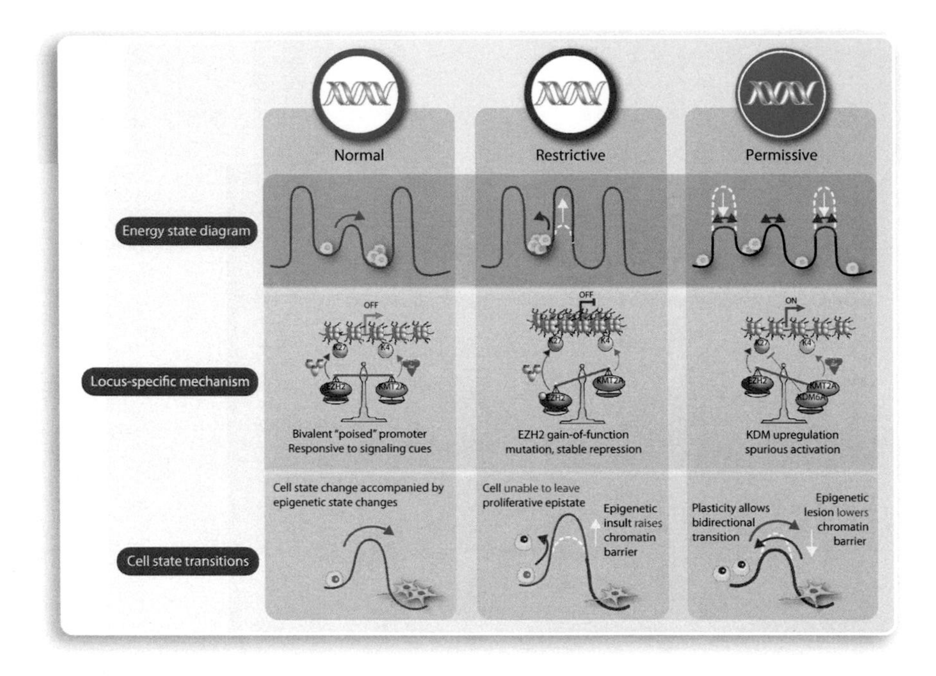

Fig. 6.6 Chromatin structure, cellular identity and cell state transitions. In normal cells (**left**) networks of chromatin proteins stabilize the states of cells but also mediate the response to intra and extracellular stimuli and occasionally allow cell state transitions. However, cells in which the chromatin network is perturbed do not respond appropriately. In restrictive chromatin (**center**) epigenetic barriers prevent cell state transitions, while in overly permissive chromatin (**right**) these barriers are lowered and allow easy transition to other cell states. The scenarios are illustrated *via* an example of the energy state change (**top**) underlying molecular mechanisms (**center**) or as cell state transitions (**bottom**). Blue nuclei represent normal cells, while red nuclei indicate cancer cells

poised constitutive heterochromatin at TSS regions. This means that their respective target genes are transcribed only in response to appropriate stimuli. In restricted cells, EZH2 may have a gain-of-function epimutation, such as often observed in several forms of lymphoma, resulting in far higher levels of repressive H3K27me3 marks, stable heterochromatin and no gene transcription. In this state, cells may be blocked in differentiation and continue to grow with a high proliferation rate. In contrast, in permissive cells a demethylase, such as KDM6A, inhibits the action of EZH2 and removes H3K27me3 marks. KDMs are often upregulated under stress conditions. In net effect, this leads to the dominance of H3K4me3 marks and to the **activation of gene expression, such as of oncogenes, even in the absence of specific stimuli**. In the cell state transition diagram (Fig. 6.6, bottom) the barrier between the cell states is either of medium height in normal cells, very high in restricted cells or rather low in permissive cells. Thus, **permissive cells can easily transform into cancer cells**.

6.4 Epigenetic Reprogramming in Cancer

Permissive chromatin has a high rate of plasticity. This allows cancer cells to acquire easily a number of different transcriptional states, e.g., shifting to alternative developmental programs, some of which are pro-oncogenic. When such an adaptive chromatin state propagates through mitosis, a new cell clone is created that overgrows other cells due to increased fitness (Fig. 6.7). This plasticity model can be considered as the epigenetic counterpart to the genetic model of genome instability being induced by carcinogen exposure or DNA repair defects. Interestingly, **in both models there are driver events, such as the activation of an oncogene, and passenger events that do not alter the fitness of the cells** (Sect. 4.4).

Chromatin homeostasis is closely linked to metabolic conditions. Many chromatin proteins, such as DNA- and histone-modifying enzymes, use metabolites as donors and co-factors, like α-ketoglutarate, the methyl donor SAM and acetyl-CoA. Therefore, they are sensitive to shifts in the concentration of these metabolites. For example, mutations of IDH enzymes lead to the accumulation of 2-hydroxyglutarate inhibiting the demethylation of DNA via TET enzymes (Fig. 6.1). The resulting DNA hypermethylation disrupts the binding of the methylation-sensitive transcription factor CTCF to insulator regions. Accordingly, the partition of the genome into discrete functional

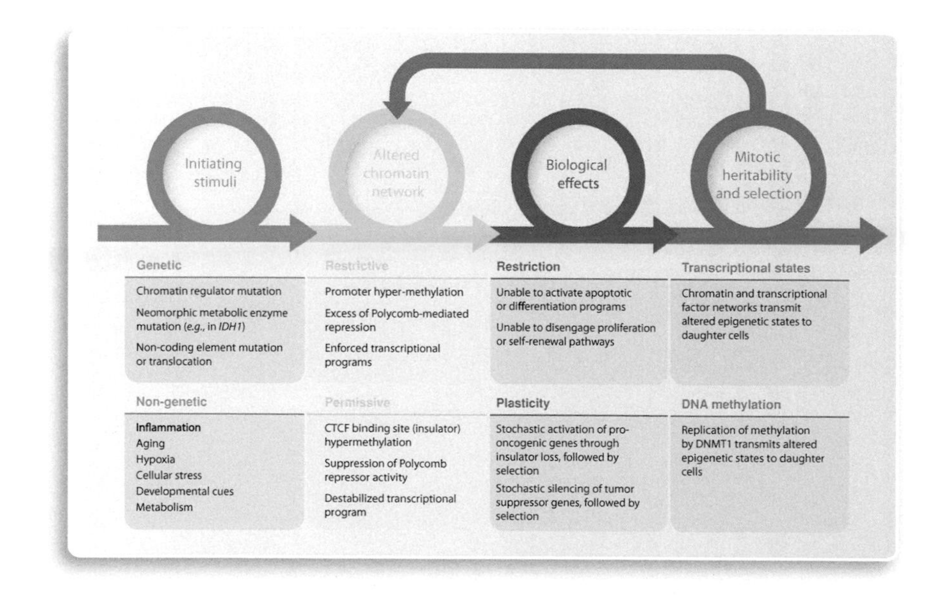

Fig. 6.7 Disruption of chromatin homeostasis in cancer. Chromatin homeostasis can be disrupted by genetic factors, such as mutations in chromatin proteins or translocation of regulatory elements, or by non-genetic factors, such as inflammation, stress or hypoxia. This can result either in very permissive or in restrictive chromatin networks. In permissive states oncogenic epigenetic changes may occur, such as silencing of tumor suppressor genes. Mitotically heritable, adaptive epigenetic changes will be selected and contribute to the hallmarks of cancer

domains, in which enhancers regulate their appropriate gene targets, is disturbed. For example, reduced CTCF binding in gliomas with *IDH* mutants leads to a transcriptome profile that indicates insulator dysfunction. Thus, **the oncogenicity of IDH mutants is primarily based on a loss of gene insulation**, i.e., on compromised CTCF-mediated genome topology, which is an effect of epigenetic dysregulation (Fig. 6.7).

The idea that cancer is fundamentally an epigenetic disease is also reflected by the relationship between cancer and the epigenetic landscape. Since epigenetic modifiers, such as genes encoding for chromatin modifiers, are highly mutated in cancer, these mutations largely affect the stability of the epigenetic landscape. (Figure 6.7). While permissive chromatin states allow oncogene activation or non-physiologic cell fate transitions, restrictive states prevent the induction of tumor suppressors or block differentiation. For example, the cancer hallmark "evasion of growth suppressors" can be based on either a loss-of-function mutation of the tumor suppressor gene *CDKN2A* or by hypermethylation of its promoter. The relative contribution of genetic and epigenetic mechanisms to the hallmarks of cancer differs between cancer types. Interestingly, the example of the adult brain tumor glioblastoma in comparison to the childhood brain tumor ependymoma suggests that long-term tumorigenesis in adults may rather be based on genetic events, while short-term tumorigenesis has majorly an epigenetic origin (Fig. 6.8). General differences between adult and pediatric cancers are summarized in Box 6.4.

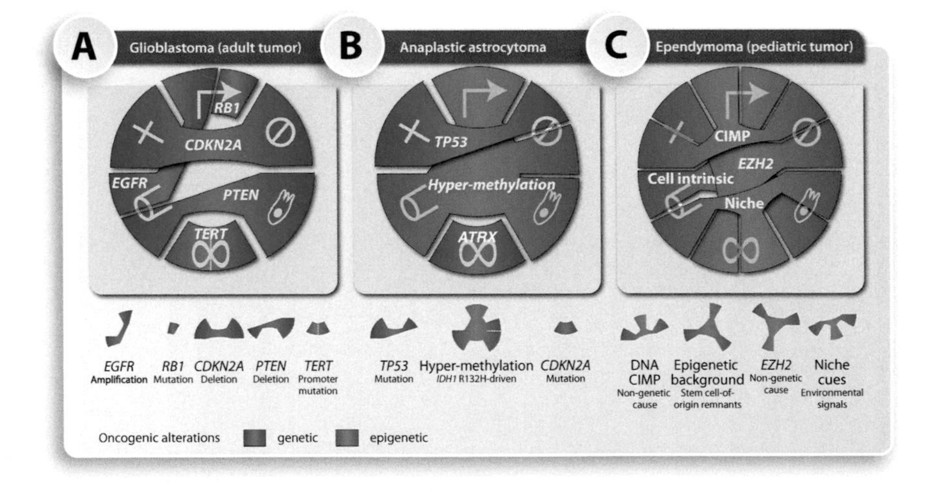

Fig. 6.8 Genetic and epigenetic mechanisms underlying the hallmarks of cancer. Both genetic (green) and epigenetic (blue) mechanisms are important factors in tumorigenesis, but their relative contribution to the hallmarks of cancer depends on the type of cancer. In the adult brain tumor glioblastoma (**A**), most hallmarks relate to genetic drivers, while in ependymoma (**C**), a childhood cancer, primarily epigenetic effects dominate. Anaplastic astrocytoma (**B**) represents an example where both genetic and epigenetic factors contribute to the hallmarks

Box 6.4: Differences between adult and pediatric malignant tumors. Cancers in adults and children cannot be compared, even if they carry the same name, such as ALL. Carcinomas, i.e., rather slowly growing tumors arising out of epithelial cells, are in adults the predominate form of cancer (Sect. 1.2), while they are rarely found in children. In general, for children cancer is a rare disease. For them malignant tumors often arise out of the hematopoietic system, such as leukemias in younger children and lymphomas in adolescence, or the brain. These cancers develop from non-epithelial cells and show a very aggressive nature combined with a high growth rate. Moreover, biological processes underlying cancer development in children differ clearly from that of adults. Tumorigenesis of adult cancers is based on the accumulation of mutations in the genome, while malignant tumors in children most often develop due to an epigenetic dysregulation during embryogenesis. Accordingly, the treatment architecture of pediatric cancers differs significantly from that of adult patients. Children tolerate far higher doses (per kg bodyweight) of chemotherapy than adults, since they have a higher stem cell division rate. This allows their healthy cells to regenerate faster from side effects of the therapy. Thus, **the cancer cure rate,** i.e.**, the 5-year disease-free survival, is much higher in children than in adults**

The known cancer genes can be classified into dominant oncogenes, which can be activated by gain-of-function mutations, amplifications or translocations (Chap. 2), and recessive tumor suppressor genes, the expression of which is often lost, like by loss-of-function mutations or methylation of their promoter regions (Chap. 3) (Table 6.1). An alternative classification is to divide all somatic mutations into drivers, the mutation of which directly affects tumorigenesis, and passengers, which are mutated as a side product, but do not have a functional contribution on oncogenesis (Sect. 4.4). The epigenetic perspective adds a further classification option for cancer genes: encoding for epigenetic modifiers, mediators and modulators. **Epigenetic modifiers** are proteins that directly modify the epigenome through DNA methylation, histone modification or structural changes of chromatin, i.e., they are chromatin modifiers and remodelers. For example, childhood cancers are often based only on a small number of genetic mutations, but these often occur in genes that encode for chromatin modifiers. Interestingly, the biallelic loss of the gene *SMARCB1* (SWI/SNF-related matrix-associated actin-dependent regulators of chromatin B1, Fig. 6.1, bottom), in pediatric rhabdoid tumors as well as in lung cancer and Burkitt's lymphoma, was a first indication that the **disruption of epigenetic control can serve as a driver for cancer**.

Epigenetic mediators are targets of epigenetic modification, i.e., they are downstream of epigenetic modifiers. For example, due to the overactivity of an epigenetic modifier, such as a chromatin modifier of the HAT family, genes encoding for pluripotency transcription factors, such as NANOG, SOX2 or OCT4 (Fig. 6.5), get activated in a somatic cell. Thus, in this situation the pluripotency transcription

Table 6.1 Three classification systems for cancer genes. Example genes for each classification are provided

Genetic classification		
Oncogene	A gene, when activated, gives advantages to cancer cells	*MYC, KRAS, PIK3CA, BRAF*
Tumor suppressor gene	A gene, when inactivated, gives advantages to cancer cells	*TP53, RB1, APC CDKN2A*
Selection classification		
Driver gene	A gene, when mutated or showing aberrant gives advantages to cancer cells	*MYC, TP53, KRAS RB1, PIK3CA*
Passenger gene	A gene, when mutated does not affect, the > 99% of all genes growth of cancer cells, i.e., not a driver	mutated in cancer
Epigenetic classification		
Epigenetic modifier	A gene that modifies the methylation of DNA, histone modification or 3D chromatin structure	*TET2, DNMT3A ARID2, BRD4 EZH2, NSD1*
Epigenetic mediator	A gene regulated by an epigenetic modifier that gives advantages to cancer cells	*OCT4, NANOG SOX2, KLF4*
Epigenetic modulator	A gene that activates or represses the the epigenetic machineries	*IDH1, CTCF*

factors are epigenetic mediators and may transform a terminally differentiated cell back to a pluripotency stage. In this way, the cell forms a cancer stem cell and initiates the tumorigenesis process. Finally, **epigenetic modulators** are gene products that are located upstream of epigenetic modifiers and mediators in signal transduction pathways. Epigenetic modulators influence the activity or localization of epigenetic modifiers in order to destabilize differentiation-specific epigenetic states. They represent a bridge between environment and epigenome. Inflammatory responses mediated by the transcription factor NF-κB are an example of an epigenetic modulator. They trigger an epigenetic switch to a positive feedback loop with the cytokine interleukin (IL) 6 and the transcription factor STAT3 in the transformation of mammary epithelia. Thus, **the actions of epigenetic modulators are often the first steps in tumorigenesis resulting in changing epigenome patterns**.

Clinical conclusion: Cancer is not only based on changes of the genome but also on epigenome-wide alterations. We cannot change mutations of our genome, but via lifestyle changes we are able to affect our epigenome and may prevent cancer. This implies the responsibility for our own health as well as a potential for a holistic treatment of disease(s).

Further Reading

Bates, S. E. (2020). Epigenetic therapies for cancer. *The New England Journal of Medicine, 383,* 650–663.

Carlberg, C., & Molnár, F. (2019). *Cancer Epigenetics* (pp. 89–99). In Human Epigenetics: How Science Works.

Corces, M. R., Granja, J. M., Shams, S., Louie, B. H., Seoane, J. A., Zhou, W., Silva, T. C., Groeneveld, C., Wong, C. K., Cho, S. W., et al. (2018). The chromatin accessibility landscape of primary human cancers. *Science, 362,* eaav1898.

Filbin, M., & Monje, M. (2019). Developmental origins and emerging therapeutic opportunities for childhood cancer. *Nature Medicine, 25,* 367–376.

Flavahan, W. A., Gaskell, E., & Bernstein, B. E. (2017). Epigenetic plasticity and the hallmarks of cancer. *Science, 357,* eaal2380.

Kinnaird, A., Zhao, S., Wellen, K. E., & Michelakis, E. D. (2016). Metabolic control of epigenetics in cancer. *Nature Reviews Cancer, 16,* 694–707.

Liu, F., Wang, L., Perna, F., & Nimer, S. D. (2016). Beyond transcription factors: how oncogenic signaling reshapes the epigenetic landscape. *Nature Reviews Cancer, 16,* 359–372.

Mohammad, H. P., Barbash, O., & Creasy, C. L. (2019). Targeting epigenetic modifications in cancer therapy: erasing the roadmap to cancer. *Nature Medicine, 25,* 403–418.

Chapter 7
Aging and Cancer

Abstract Aging is the progressive decline in the function of cells, tissues and organs that leads to impaired functions of our body. Accordingly, older age is the primary risk factor for non-communicable diseases, such as cancer. Common hallmarks of aging are cellular senescence, genome instability, epigenetic alterations and telomere attrition. Interestingly, epigenetic signatures can serve as biomarkers of aging, since the process is associated with specific chromatin patterns and chromatin modifiers that are able to modulate both life- and healthspan. The shortening of telomeres has on one hand a tumor-suppressive effect, since it induces cell cycle arrest, but on the other hand it can also lead to extensive genome instability that promotes cancer progression.

Keywords Lifespan · Healthspan · Cellular senescence · Hallmarks of aging · Epigenetic biomarkers · Telomere shortening · Telomere crisis

7.1 Central Role of Aging During Chronic Diseases

Our average life expectancy has drastically increased during the past 200 years showing a rise of approximately 2.5 years per decade. This boost is mainly due to changes in our living conditions, such as improved food, water and hygiene, the reduced impact of infectious diseases via vaccinations, the use of antibiotics and a better medical care in general. Nowadays, we live beyond the ages at which in the past most of us would have been already dead. **This gives cancer at present a far higher impact as a cause of death than it had in the past**.

There are no evolutionary mechanisms that are able to select for maintaining fitness at older ages. Therefore, older age is the major risk factor for diverse types of loss of physiological functions of our body and for a high prevalence for non-communicable chronic diseases, such as cardiovascular disease, atherosclerosis, type 2 diabetes, stroke, dementia and neurodegeneration, sarcopenia, frailty, osteoporosis, arthritis, renal failure, macular degeneration and, of course, all types of cancer (Fig. 7.1). In this perspective, cancer parallels many other chronic non-communicable diseases, the incidence of which significantly rose during the past 100 years due to increased life expectancy. Unfortunately, our **healthspan**, i.e., the

C. Carlberg and E. Velleuer, *Cancer Biology: How Science Works*,
https://doi.org/10.1007/978-3-030-75699-4_7

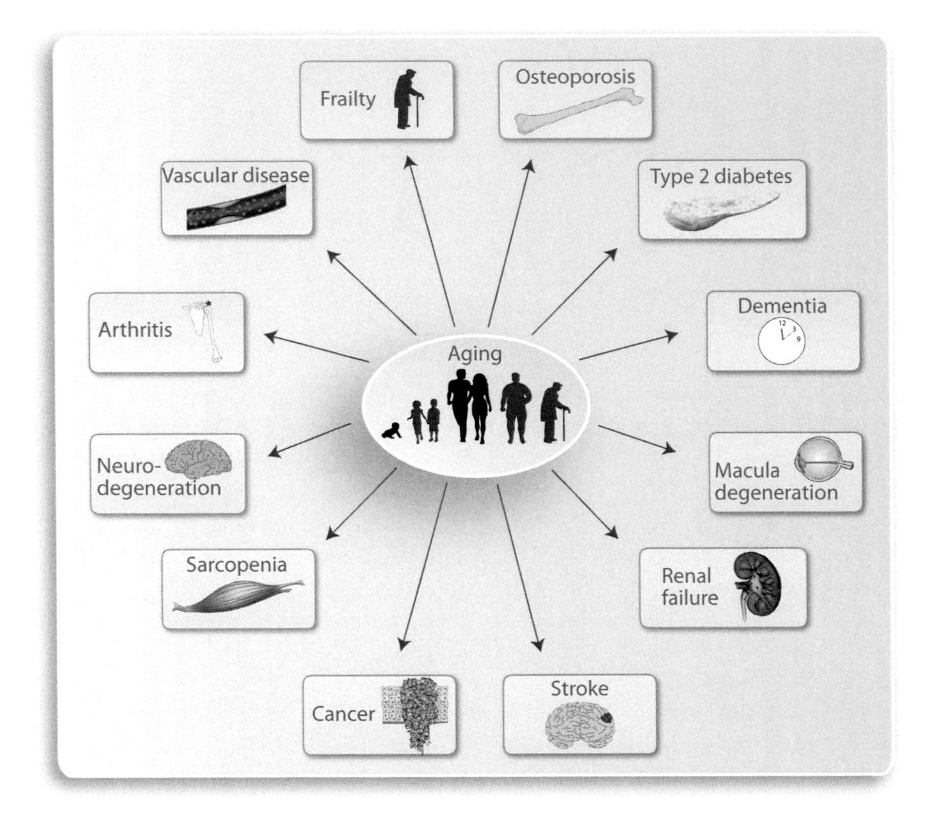

Fig. 7.1 Aging and non-communicable chronic diseases. Aging is the major risk factor for all listed chronic non-communicable diseases

disease-free period of our life, has not increased with the same speed as our **lifespan**. This implies that we spend in average more time, mostly during the last decades of our life, with suffering from chronic diseases. Therefore, there is an urgent need to reduce chronic diseases in the elderly to a minimum by applying preventive interventions and stimulating healthy lifestyle choices (Sect. 1.5). Ideally, **this should restrict morbidity to the very end of life**.

The maximal lifespan of humans is in the order of 110-120 years, but is reached only by a very few individuals. This makes us the longest living mammal on land (some whale species live even up to 200 years), i.e., our speed of aging is slower than for most other species. Elephants have with a maximal lifespan of 65 years the second longest life of land-living mammals. In contrast to humans, elephants have since millions of years a high life expectancy, since they have no natural predators and their living conditions did not change significantly. Nevertheless, despite their high age and huge body size, elephants rarely get cancer. This seems to be related to the fact that elephants have 20 copies of *TP53* and related genes in their genome, i.e., that they are better protected to a loss of p53 activity than humans. Since p53 is a strong inducer of senescence (Box 7.1) a balance between the protection of cancer and the induction of physiological aging has to be found. Thus, **humans live longer than elephants but get far more likely cancer**.

Box 7.1: Senescence. Cellular senescence is a stable arrest of the cell cycle coupled with a drastic change of their secretome (i.e., the complete set of secreted molecules), which includes pro-inflammatory cytokines. Senescence is a compensatory response that is beneficial for the body, since it leads to the elimination of damaged and potentially oncogenic cells. However, this requires that senescent cells are efficiently cleared and replaced. In the original description by Hayflick, senescence is caused by telomere shortening (Sect. 7.4). Additionally, also other forms of DNA damage, which are coupled with the activation of tumor suppressors, such as p16, p19 and p53 (Sect. 3.2), can trigger the process. Senescence occurs in all phases of life, but in older age the clearance of senescent cells by the immune system (Chap. 10) is less efficient. Thus, **senescent cells accumulate with aging not only due to increased DNA damage but also because of decreased immune surveillance**

Human lifespan comprises a period of growth and differentiation that ends up in sexual maturity, i.e., in a period of maximal fitness and fertility, and a period of aging that comes with loss-of-function at the various levels of cells, tissues and the organism as a whole (Fig. 7.2). Due to a rather long time until sexual maturity (12-15 years) and some 20-25 years of childcare, evolution selected us to be for some 40 years mainly free of non-communicable diseases, such as cancer. For comparison, the longest-lived vertebrate, the Greenland shark, needs 150 years to reach sexual maturity and lives some 400 years. Thus, **lifespan seems to be proportional to the onset of maturity**.

We humans seem to have a kind of "warranty" to get at least some 40 years old, in order to guarantee the survival of our offspring. However, after this age the rate of cancer is drastically increasing. Interestingly, the curve of increase of cancer incidence mirrors the curve of decrease in physiological fitness (Fig. 7.2). **The remaining 80 years up to the maximal lifespan of 120 years can be considered as security measure.** However, this latter phase is not under evolutionary control and therefore associated with a wide range of non-communicable diseases.

7.2 The Hallmarks of Aging

Aging is defined as a time-dependent function decline of our body due to the accumulation of cellular damage. In this mechanism aging and tumorigenesis show similarities. However, while in the aging process damaged cells behave altruistically and

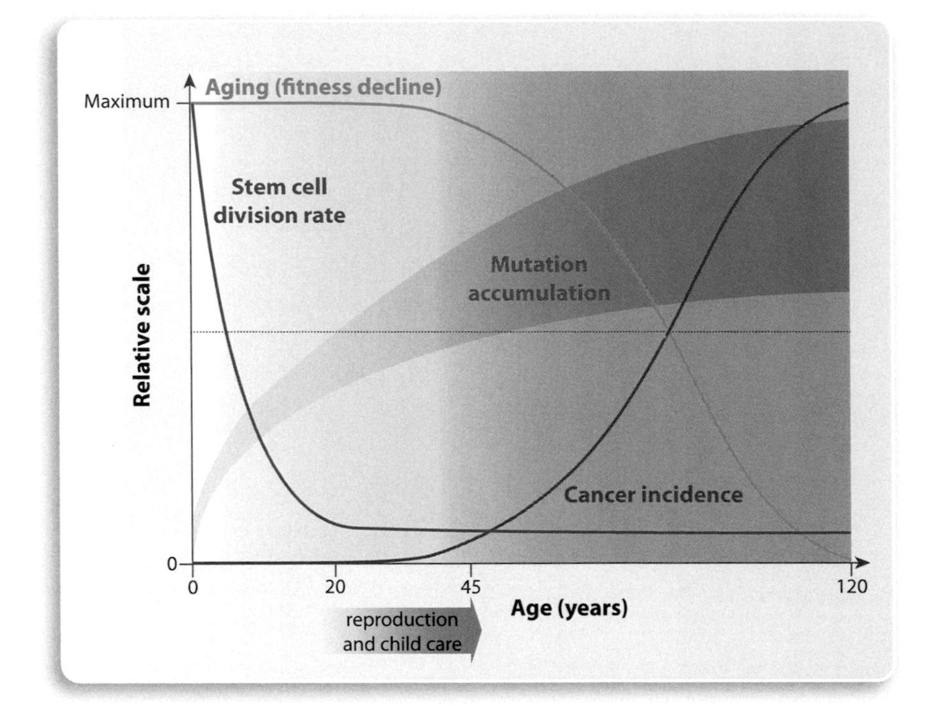

Fig. 7.2 Aging, mutations and cancer. With increasing age stem cell division rate (based on hematopoietic stem cells) declines as well as the overall fitness many decades later. Similarly, an accumulation of mutations is followed after decades by an increase of cancer incidence. The latter mirrors the curve of physiological decline

shift into senescence for the benefit of the whole body, in tumorigenesis cells act egoistically and take advantage from the alteration of their genome for growing into malignant cells that harm the body. Thus, **aging and cancer can be considered as two different expressions of the same underlying mechanism.**

There are nine hallmarks that contribute to aging and determine its phenotype (Fig. 7.3). Genome instability, telomere attrition, epigenetic alteration and loss of proteostasis are hallmarks that create damage, i.e., they are considered as **primary causes of aging**. In contrast, the three hallmarks deregulated of nutrient sensing, mitochondrial dysfunction and cellular senescence are **antagonistic responses to damage**. Finally, the two hallmarks stem cell exhaustion and altered intercellular communication integrate the negative consequences of the aging phenotype and are responsible for the **functional decline**.

Genome instability is a common hallmark of aging and cancer. It represents the accumulation of DNA damage in form of SNVs and genome rearrangements (Sect.4.3). The accumulation of DNA mutations or even aneuploidy (chromosome gains and losses), such as during aging, has an effect on both the transcriptome and the epigenome. This damage affects the function of essential intracellular signaling

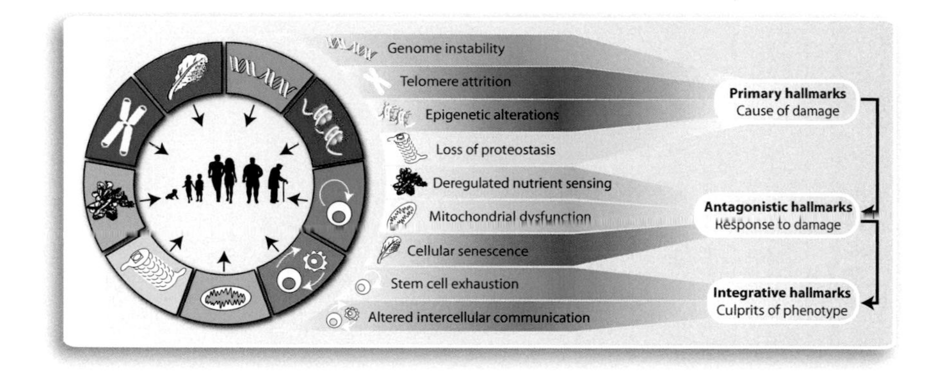

Fig. 7.3 Hallmarks of aging. Nine major hallmarks of aging are listed and grouped into the three categories (i) primary causes of cellular damage, (ii) compensatory or antagonistic responses to the damage and (iii) responsible for the functional decline

pathways and causes dysfunction of the cell. In case these damaged cells are not eliminated by either apoptosis or senescence, they risk to disturb tissue homeostasis, in particular, when the damage concerns stem cells. Interestingly, a number of pre-mature aging syndromes (Box 7.2) are monogenetic diseases that have in common a defect in a DNA repair pathway and are coupled with early onset of cancer, such as in Fanconi anemia and Xeroderma pigmentosum. Thus, **the accumulation of DNA damage in young age due to a defective DNA repair system accelerates the aging process**. The primary aging hallmarks **epigenetic alterations** and **telomere attrition** will be discussed in Sects. 7.3 and 7.4. Proteostasis comprises both the stabilization of correctly folded proteins as well as the degradation of damaged or misfolded proteins by the autophagy-lysosomal system and the ubiquitin-proteasome system. Thus, **loss of proteostasis** is another primary aging hallmark. It describes that the aging process is stimulated by the loss of control mechanisms that preserves the stability and functionality of proteins. Importantly, the accumulation of unfolded, misfolded or aggregated proteins is a common mechanism of Alzheimer's disease and other neurodegenerative disorders.

Box 7.2: Diseases with defective DNA repair and pre-mature aging are cancer predisposition syndromes. Deficiencies in DNA repair pathways are the basis of a number of progeroid syndromes, such as Werner syndrome, Ataxia telangiectasia, Bloom syndrome, Nijmegen breakage syndrome, Xeroderma pigmentosum, Dyskeratosis congenita and Fanconi anemia. Some genes underlying these diseases are known tumor suppressors, such as *ATM*, *BRCA1*, *BRCA2*, *BRIP1* and *PALB2* (Fig. 5.1). The impaired DNA damage control in these syndromes leads to chromosomal aberrations and genome instability. Clinically, most of the patients present with the triad

- congenital anomalies including growth retardation
- failures in tissues with high cell division rates, such as the bone marrow
- a high risk to develop cancer.

Genome instability leads to significantly reduced stem cell division rate, so that the patients face already at young age age-related disease, such as glucose intolerance, early menopause, osteoporosis, renal failure and cancer (Fig. 7.1). Most of these cancers concern the hematopoietic system and the epithelia compartment, such as myelodysplastic syndrome and squamous cell carcinoma

Deregulated nutrient sensing belongs to the hallmarks that represent a response to the primary hallmarks of aging. Accordingly, decreased nutrient signaling, such as induced by dietary restriction, extends longevity, while anabolic signaling accelerates aging. Furthermore, **mitochondrial dysfunction** is one of the aging hallmarks trying to counteract to the effects of the primary hallmarks. Mitochondria are the power plants of our cells and their dysfunction via increased electron leakage and reduced ATP generation accelerates aging. The latter reduces the efficiency of mitochondrial bioenergetics and induces emergency programs, such as apoptosis (Box 3.2). In addition, during aging there is reduced turnover of mitochondria via mitophagy. The third antagonistic hallmark, the induction of **cellular senescence**, was already discussed in Sect. 7.1. At low level, antagonistic hallmarks mediate beneficial effects but become deleterious at high levels.

Stem cell exhaustion is an integrative consequence of various aging-related cell damages and leads to the decline of the regenerative potential of tissues. For example, the potency of hematopoiesis drops with age and leads to a reduced function of the immune system. In turn, the rejuvenation of stem cells may reverse some aspects of the aging process. Finally, aging leads also to **altered intercellular communication** on endocrine, neuroendocrine and neuronal level. A prominent aging-associated process is **inflammaging**, a pro-inflammatory phenotype that parallels with immunosenescence, which is a decline of the immune system. **Immunosenescence worsens the aging process on the whole body level, since the immune system fails to eliminate infectious agents, infected cells and transformed cells.** Both integrative hallmarks of aging affect tissue function and homeostasis. They occur primarily, when the accumulated damage caused by the primary hallmarks cannot be compensated by the antagonistic hallmarks.

7.3 Epigenetics of Aging

Cells of young individuals show a robust transcriptome and normal chromatin states, but with increasing age the transcriptome gets instable and aberrant chromatin states are accumulating, i.e., **epigenetic alterations are a hallmark of aging**. For example, DNA damage stimulates the recruitment of chromatin modifiers that may induce abnormal chromatin states. In turn, epigenome-wide changes during aging can increase the susceptibility of the genome to mutations and in parallel reduce the precision of transcription. Moreover, errors in DNA repair and failure to correctly replicate the genome and epigenome not only increase the number of DNA mutations but also of epimutations. Thus, since genome surveillance and epigenetic remodeling influence each other, **environment-induced epigenome instability throughout life is an important driver of the aging process**.

The epigenome is able to preserve the results of cellular perturbations by environmental factors in form of changes in DNA methylation, histone modifications and 3D organization of chromatin. Changes in epigenomic patterns are **epigenetic drifts** that describe the lifelong information recording, i.e., a kind of memory, of somatic cell types and tissues. Epigenetic drifts, such as hypermethylation of CpG islands close to the regulatory regions of tumor suppressor genes, contribute to the risk for cancer (Sect. 6.2). Changes of the epigenome, in particular the DNA methylome, are associated with the chronological age of an individual as well as with age-related diseases, such as cancer. The DNA methylation status at a few hundred key CpG islands, which can be measured from easily accessible tissues and cell types, such as skin or peripheral blood mononuclear cells, is used as a biomarker. The respective chromatin landscapes are then correlated with chronological age and biological age, i.e., the age at which the population average is most similar to the individual. DNA methylome profiling of a large cohort of individuals spanning over a wide age range provides a good correlation between chronological and biological age, but there are also significant interindividual variations. At a given chronological age the investigated tissue of some individuals has a far "younger" epigenome, while that of others is already "older" (Fig. 7.4). Accordingly, it can be expected that the latter individuals may have an earlier onset of age-related diseases, such as cancer, from the respective tissue and eventually may die at younger chronological age than the former individuals. This has been observed with individuals suffering from pre-mature aging syndromes (Box 7.2). In contrast, the blood of the offspring of super-centenarians, i.e., individuals who reached an age of at least 105 years, has a lower epigenetic age than that of age-matched controls. Thus, **epigenetic signatures can serve as biomarkers of aging**. They may be druggable targets, in order to delay or even reverse age-related diseases, such as cancer (Sect. 11.2).

In addition to changes in DNA methylation the key epigenetic hallmarks of aging (Fig. 7.5) include

- a general loss of histones due to local and global chromatin remodeling
- an imbalance of activating and repressive histone modifications
- site-specific loss and gain in heterochromatin

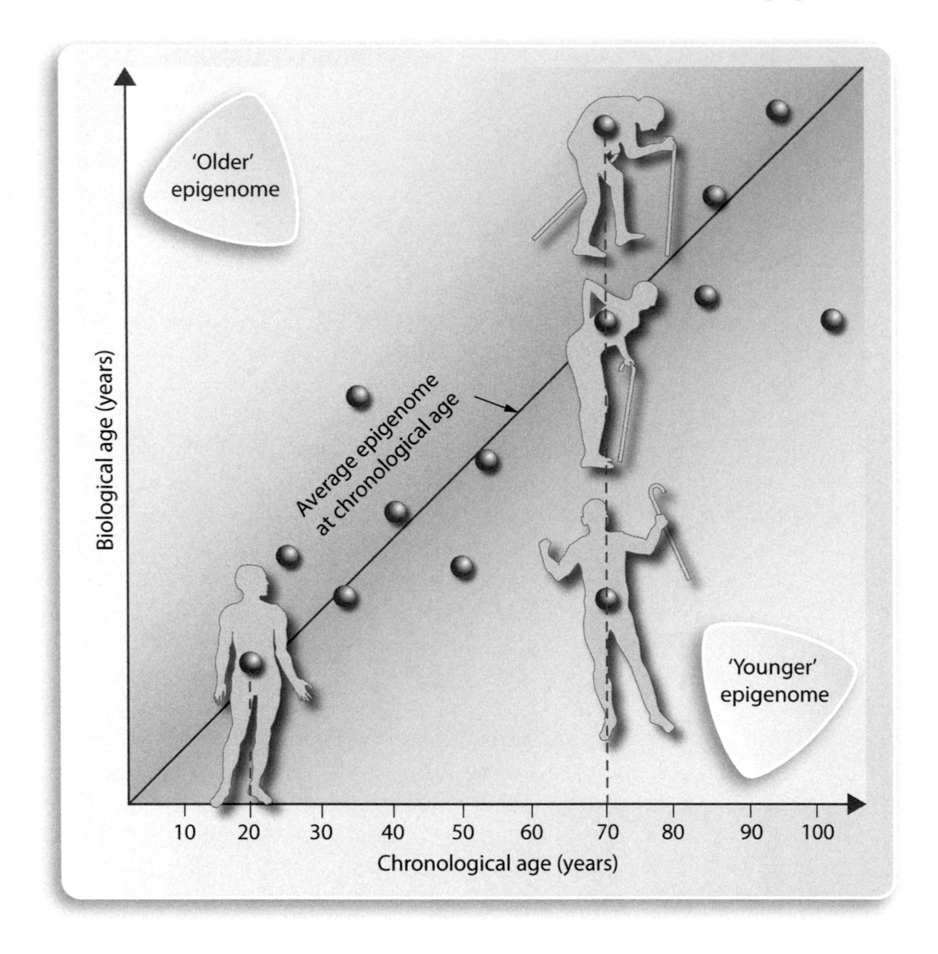

Fig. 7.4 Epigenetic biomarkers of age. Epigenome-wide patterns not only monitor cellular identities but also cellular health and age. For example, changes in the DNA methylation status of CpG islands, as measured in white blood cells, taken from individuals of different age (balls), can serve as a sensor for chronological age. However, there is significant deviation from the postulated linear fit (diagonal line) suggesting that methylation patterns also represent biological age

- significant nuclear reorganization
- transcriptional changes.

The general loss of histones in aging cells is tightly linked to cell division (Fig. 7.5A). Cells then develop senescence-associated heterochromatin foci, which are regions of highly condensed chromatin associated with heterochromatic histone modifications, heterochromatic proteins and the histone variant macroH2A. In general, on some 30% of the genome of senescent cells chromatin is reorganized with an increase of H3K4me3 and H3K27me3 marks within LADs and a loss of H3K27me3 outside of LADs (Fig. 7.5B).

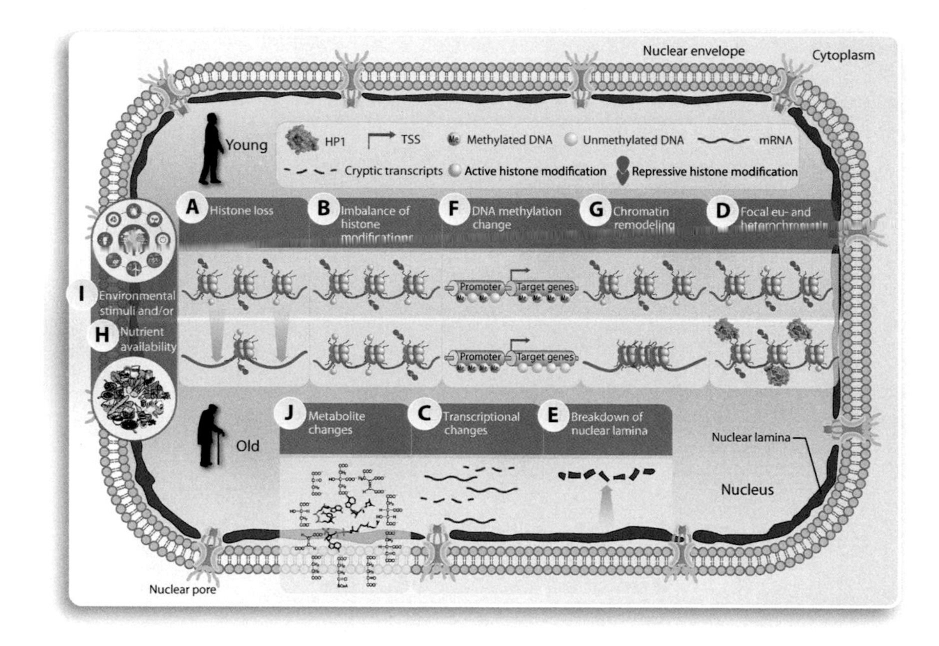

Fig. 7.5 Epigenetics of senescence and aging. The epigenetic hallmarks of senescence and aging are a loss of histones (**A**), imbalance of activating and repressive modifications (**B**), changes in gene expression (**C**), losses and gains in heterochromatin (**D**), breakdown of nuclear lamina (**E**), global hypo-methylation and focal hypermethylation (**F**) as well as chromatin remodeling (**G**). These changes are heavily dictated by environmental stimuli (**H**) and nutrient availability (**I**) that in turn alter intracellular metabolite concentrations (**J**)

The most significant molecular consequence of the loss of repressive histone marks and the gain of activating marks during aging is a change in gene expression, such as the upregulation of genes related to cell adhesion and ribosomal proteins, while genes related to cell cycle regulation, DNA repair and DNA replication are downregulated (Fig. 7.5C). The lack of some of the repressed genes negatively affects longevity, while some of the activated genes are detrimental to lifespan. Moreover, the transcriptional reprogramming happening during the onset of senescence also occurs through altered activity of chromatin modifiers and remodelers. Constitutive heterochromatin at telomeres, centromeres and peri-centromeres is established during embryogenesis and is thought to be maintained throughout lifespan. However, senescent cells loose some of these regions of constitutive chromatin, resulting in an increase of euchromatic regions (Fig. 7.5D). Moreover, a loss of the nuclear lamina (Fig. 7.5E) stimulates the breakdown of heterochromatin organization and the re-localization of heterochromatic proteins to regions in the genome where they contribute to the formation of region-specific foci. The global hypomethylation and local hypermethylation of genomic DNA during aging fits with the observation of global heterochromatin deregulation in combination with focal increase at some genomic regions (Fig. 7.5F). DNA methylation is primarily lost at repetitive genomic regions that are in constitutive heterochromatin, while hypermethylation mostly occurs at CpG islands close to promoter regions. SWI/SNF chromatin remodelers are associated with gene activation and they seem to promote aging, while repressive chromatin remodelers support longevity (Fig. 7.5G).

Evolutionary highly conserved proteins, such as insulin and IGF receptors, the amino acid sensor mTOR and the nicotinamide adenine dinucleotide (NAD^+)-sensing HDAC sirtuin 1 (SIRT1), belong to the nutrient signaling pathways that integrate metabolic signals into chromatin responses. They inform the epigenome on nutrient availability and their misregulation is one of the hallmarks or aging (Fig. 7.5H). Similar principles apply to the sensors of other environmental inputs (Fig. 7.5I) or intracellular metabolites (Fig. 7.5 J), such as steroid hormones like estrogen and testosterone via their nuclear receptors ESR1 and AR or α-ketoglutarate via TET enzymes, that affect longevity via changes in the chromatin landscape.

7.4 Telomeres and Replicative Immortality

Although DNA damage accumulates with age in all regions of the genome in a random fashion, telomeres show a particular susceptibility to aging. Telomeres are repetitive DNA sequences (mainly TTAGGG repeats) of approximately 12 kb in length that are found at the end of all chromosomes. Chromosome ends have the risk to be recognized by the NHEJ DNA repair pathway as a double-strand break, which may lead to chromosome fusions. Therefore, sheltering protein complexes at telomeres represses the activation of this DDR by inhibiting the DNA damage sensing kinase ATM and ATR and preventing the activation of p53 (Sect. 3.1). Moreover, during replication the ends of linear DNA, such as at telomeres, are a challenge for DNA polymerases to be copied in completeness. Thus, without the help of the telomerase enzyme complex after every cell division some 50-100 bp of telomere sequence would be lost (Fig. 7.6, top). The telomerase enzyme acts as a reverse transcriptase that synthesizes telomeric DNA de novo using integral *TERC* (telomerase RNA component) RNA as the template and the 3'-end of the chromosome as the primer. In normal somatic cells the activity of the telomerase is downregulated via the silencing of the *TERT* gene, which encodes for the reverse transcriptase subunit of the enzyme complex. Thus, **with the exception of some stem cells, in each round of replication telomeres get shorter**.

Telomere attrition is one of the primary hallmarks of aging (Sect. 7.2). When multiple cell divisions have reduced the telomere length to some 4 kb in at least a few chromosomes, telomere crisis occurs, which activates a DDR via p16, p21 and p53 arresting the cell cycle and inducing apoptosis or senescence (Fig. 7.6, center). This proliferation barrier applies also to malignant tumor cells that have not managed to reactivate the *TERT* gene, i.e., **telomere shortening protects against the development of malignant tumors**. Interestingly, tumor suppression via telomere

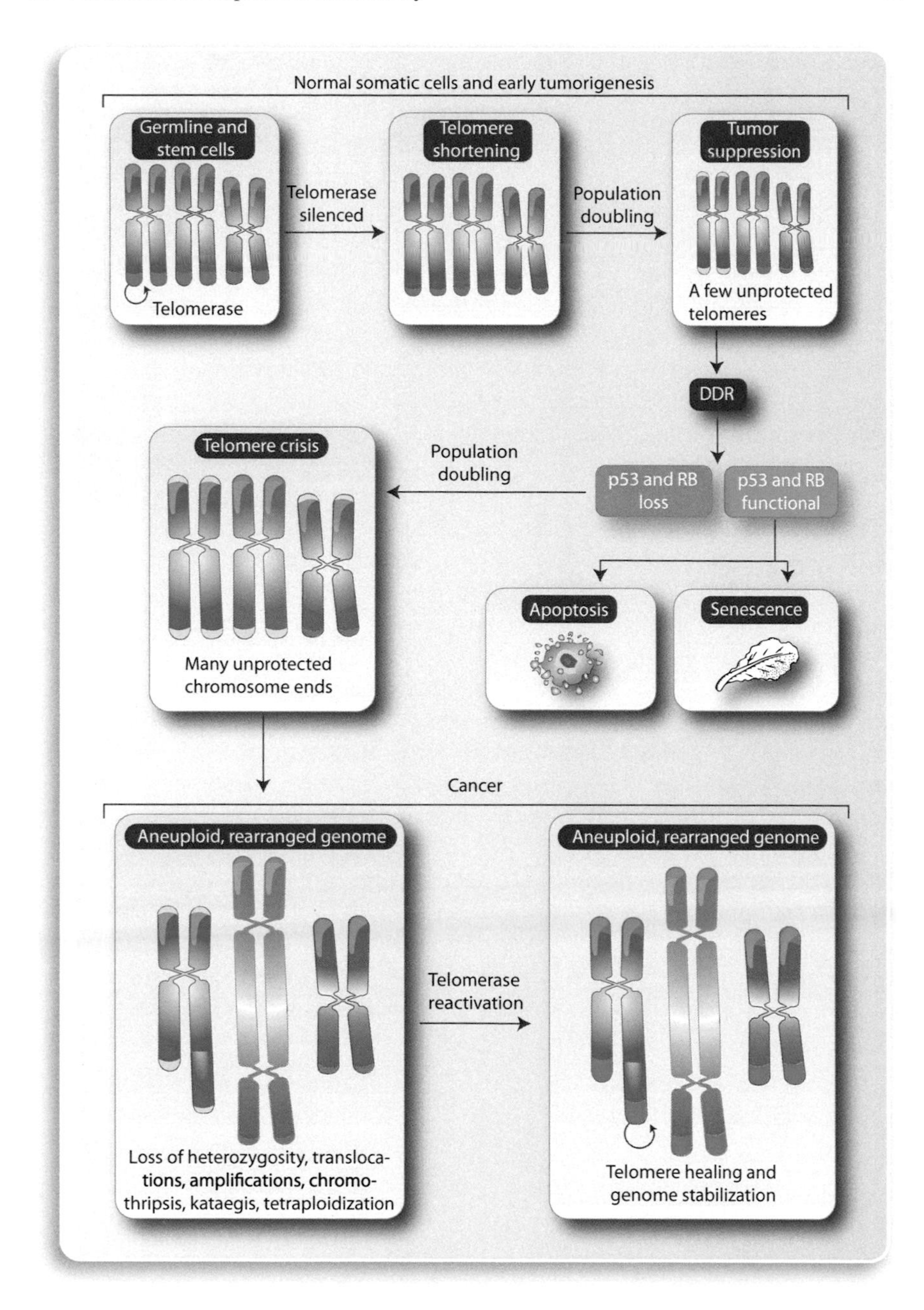

Fig. 7.6 Impact of telomere attrition. Since in normal and early cancer cells there is no active telomerase due to a silenced *TERT* gene, gradual telomere attrition happens (**top**). When after a number of DNA replication rounds a few telomeres become too short, DDR signaling *via* p53 and RB induces cell cycle arrest leading either to apoptosis or senescence (**center right**). Loss of RB and p53 tumor suppressor pathways disables this response and when the cells continue to divide, further shortening of their telomeres induces a telomere crisis (**center**). Then unprotected chromosome ends generate end-to-end fusions and dicentric chromosomes, leading to many forms of genome instability leading to cancer (**bottom**). Finally, telomerase reactivation solves the telomere crisis and heals critically shortened telomeres, giving cells with a rearranged genome growth advantage

attrition is restricted to large animals with a reproductive strategy that requires a long lifespan, like humans and elephants, but not to mice.

Cancer cells can only escape from telomere crisis when they manage to express the *TERT* gene (Fig. 7.6, bottom). TERT protein then adds GGTTAG repeats to the 3'-terminus of chromosomal DNA. This restores the telomere function and provides the capacity for unlimited proliferation, i.e., the cancer cells are not driven into senescence. Accordingly, the obtained **replicative immortality is one of the six key hallmarks of cancer** (Sect. 2.4). Some 90% of human cancer cells escape telomere crisis by *TERT* activation, while the remaining use the "alternative lengthening of telomeres" pathway, in which telomeres are lengthened through the HR DNA repair system that is induced by loss-of-function mutations of the genes *ATRX* (ATRX chromatin remodeler) and *DAXX* (death domain associated protein). Most telomerase-positive cancer cells do not only upregulate the expression of the *TERT* gene but carry in addition an extensive amount of genome rearrangements, such as deletions (leading to loss of the gene), amplifications, translocations, chromothripsis and kataegis (Sect. 5.3). Furthermore, cancer cells that have lost a functional p53 or RB tumor suppressor pathway may pass the senescence barrier and further shorten their telomeres. These dysfunctional telomeres have a high chance that the NHEJ repair pathway fuses them to each other and creates end-to-end fused dicentric chromosomes, often leading to tetraploidization (doubling the set of chromosomes) and increases genome instability. This further boosts the mutability of the genome and accelerates driver mutations in oncogenes and tumor suppressor genes. This demonstrates again that **genome instability is both a hallmark of aging and cancer**. Overcoming the telomere crisis is on one hand stabilizing cancer cells concerning unlimited proliferation, but one the other hand their genome is getting more and more unstable.

Clinical conclusion: The processes aging and tumorigenesis are closely related via their common basis "accumulation of genome instability". The cancer prevention mechanism "induction of senescence" leads over time, in particular in combination with decreased competence of the immune system, to functional deterioration of our body. Thus, our body has to choose between cancer and aging.

Further Reading

Gorgoulis, V., Adams, P. D., Alimonti, A., Bennett, D. C., Bischof, O., Bishop, C., et al. (2019). Cellular senescence: defining a path forward. *Cell, 179,* 813–827.

He, S., & Sharpless, N. E. (2017). Senescence in health and disease. *Cell, 169,* 1000–1011.

Jaiswal, S., & Ebert, B. L. (2019). Clonal hematopoiesis in human aging and disease. *Science, 366*, aan4673.

Khosla, S., Farr, J. N., Tchkonia, T., & Kirkland, J. L. (2020). The role of cellular senescence in ageing and endocrine disease. *Nature Reviews Endocrinology, 16,* 263–275.

Lopez-Otin, C., Blasco, M. A., Partridge, L., Serrano, M., & Kroemer, G. (2013). The hallmarks of aging. *Cell, 153,* 1194–1217.

Maciejowski, J., & de Lange, T. (2017). Telomeres in cancer: tumor suppression and genome instability. *Nature Reviews Molecular Cell Biology, 18,* 175–186.
Shay, J. W., & Wright, W. E. (2019). Telomeres and telomerase: three decades of progress. *Nature Reviews Genetics, 20,* 299–309.

Chapter 8
Tumor Microenvironment

Abstract The tumor microenvironment is a stroma of normal cells supporting cancer cells. Tumors act like parasites that use the physiological wound healing response of the host to acquire a stroma that they need for survival and growth. The chronic inflammatory setup of tumors let them behave as wounds that do not heal. In concert with tumor-associated stromal cells, such as fibroblasts and a variety of immune cells, cancer cells start the process of angiogenesis that is essential for their support with oxygen and nutrients. Moreover, TAMs constitute a major component of the tumor microenvironment and shape the overall metabolic profile of the tumor microenvironment supporting disease progression. Thus, this chapter discusses the hallmarks of cancer "inducing angiogenesis, "tumor-promoting inflammation" and "deregulation of cellular energetics".

Keywords Wound healing · Chronic inflammation · Angiogenesis · Tumor-associated macrophages · Metabolism

8.1 The Impact of the Wound Healing Program for Cancer

Normal healthy tissues are formed by two major compartments: **parenchyma**, i.e., epithelium, and **stroma**. The latter is composed of:

- cells with fixed positions, such as fibroblasts, that provide together with extracellular matrix proteins (Box 8.1) physical stability to the tissue
- blood vessels made out of endothelial cells that transport oxygen, nutrients and clear waste metabolites to/from the tissue
- mobile immune cells, such as myeloid cells and lymphocytes, that mediate tissue defense.

Tumors have the same needs compared to healthy tissues and are similarly organized, where parenchyma represents malignant cells and stroma the supporting **microenvironment** that is formed by non-malignant cells. Paradoxically, normal cells of the microenvironment support the tumor with oxygen and nutrients, i.e.,

C. Carlberg and E. Velleuer, *Cancer Biology: How Science Works*,
https://doi.org/10.1007/978-3-030-75699-4_8

cancer cells act like parasites behaving in a suicidal manner, since they may eventually kill their host and thus also themselves. Solid malignant tumors, such as carcinomas and sarcomas, induce **angiogenesis** in their microenvironment, since they need to the vascularized (Sect. 8.3). The latter process largely depends on the overexpression of growth factors of the VEGF family that increase vascular permeability to plasma and initiate a cascade of molecular and cellular events resembling that of a healing wound.

Box 8.1: Extracellular matrix. The extracellular matrix is a 3D network of extracellular macromolecules, such as collagen, enzymes and glycoproteins, that provides structural and biochemical support of cells in their environment. The composition of the extracellular matrix varies between tissues but cell adhesion, cell-to-cell communication and differentiation are common functions. The interstitial matrix and the basement membrane are components of the extracellular matrix.

- the interstitial matrix is the space between cells. It is filled with polysaccharides and fibrous proteins and buffers against mechanical stress of the extracellular matrix
- the basement membrane is formed of sheet-like depositions between epithelial tissues and the underlying connective tissue, i.e., it is a membrane on which epithelial cells rest.

The extracellular matrix is a critical regulator of cancer cell growth. It has to be modelled in its composition, e.g., by matrix metalloproteinases (MMPs) produced by cells of the tumor microenvironment, in order to allow invasion and metastasis or angiogenesis. In malignant tumors the composition and topography of the vascular and interstitial extracellular matrix are altered supporting its pro-angiogenic and vascular-stabilizing function

Normal **wound healing** (Box 8.2) can be subdivided into the phases hemostasis, inflammation, proliferation, angiogenesis and generation of mature connective tissue stroma. However, in the context of some diseases, such as foot ulcer, wounds heal very slowly or not at all. In these non-healing wounds, the inflammatory phase is chronic, i.e., instead of a week it lasts months or years. In these chronic wounds there is a competition between pro- and anti-inflammatory signals, which leads to an environment that does not allow proper healing. Interestingly, the formation of tumor stroma follows the same principles, i.e., tumors use regular cellular processes for their support but do not stop their demands as a healed wound would do. Thus, **tumors behave like non-healing wounds**.

Box 8.2: Wound repair. Wound healing is a complex dynamic process that starts within the first few minutes after injury with **hemostasis**, i.e., blood coagulation, where the protein thrombin induces a conformational change to platelets from the blood that allows clotting (Fig. 8.1). Platelets release chemokines and other signals to promote clotting via the activation of the protein fibrin that forms a mesh sticking the platelets to each other. The resulting clot formed of fibrin, fibronectin, vitronectin and thrombospondin serves as an insoluble plug that prevents further bleeding. During the **inflammation** phase damaged and dead cells and microbes are cleared via phagocytosis by macrophages and neutrophils, i.e., cells of the innate immune system. This immune response is initiated by **damage-associated molecular patterns (DAMPs)** released by damaged and dead cells and **pathogen-associated molecular patterns (PAMPs)** originating from microbes. Neutrophils are attracted by pro-inflammatory signals, such as IL1β, tumor necrosis factor (TNF) and bacterial endotoxins like lipopolysaccharide (LPS). The activation of inflammatory signaling pathways, such as mediated by the transcription factor NF-κB, leads to a cascade of cytokine release from neutrophils and other stroma cells. Like neutrophils, also monocytes enter the wound tissue from the blood stream, where they differentiate into macrophages that act as master effector cells in tissue repair. Platelet-derived growth factors (PDGFs) and other pro-inflammatory signals released by platelets and immune cells coordinate these cellular migration and division during the **proliferation** phase. In this phase, angiogenesis, collagen deposition, granulation tissue formation, epithelialization and wound contraction occur. In **angiogenesis**, vascular endothelial cells form new blood vessels, while fibroblasts proliferate and secrete proteins, such as collagen and fibronectin, that form a new extracellular matrix. In parallel, epithelial cells undergo **EMT** (Fig. 9.2), which allows them to proliferate and change their morphology for "crawling" on top of the wound bed. Once the wound is completely covered, the cells undergo the opposite process, referred to as **mesenchymal-epithelial transition (MET)**, i.e., they differentiate back to regular epithelial cells and stop proliferating. Finally, myofibroblasts grip the wound edges and contract, in order to decrease the size of the wound. Moreover, during this maturation phase collagen is realigned along tension lines, and cells that are no longer needed are removed by apoptosis

An important feature of wound healing is **EMT** of epithelial cells, such as keratinocytes of wounded skin, i.e., normal terminally differentiated cells are able to activate a program that let them start to proliferate, changes their morphology and allows them to move to a distant site. The same process is activated in metastatic cancer cells when they evade the primary malignant tumor tissue, i.e., the ability to perform metastasis is a property of normal cells that is misused by malignant cells (Sect. 9.1). The migration of cells is facilitated by the release of MMPs that resolve

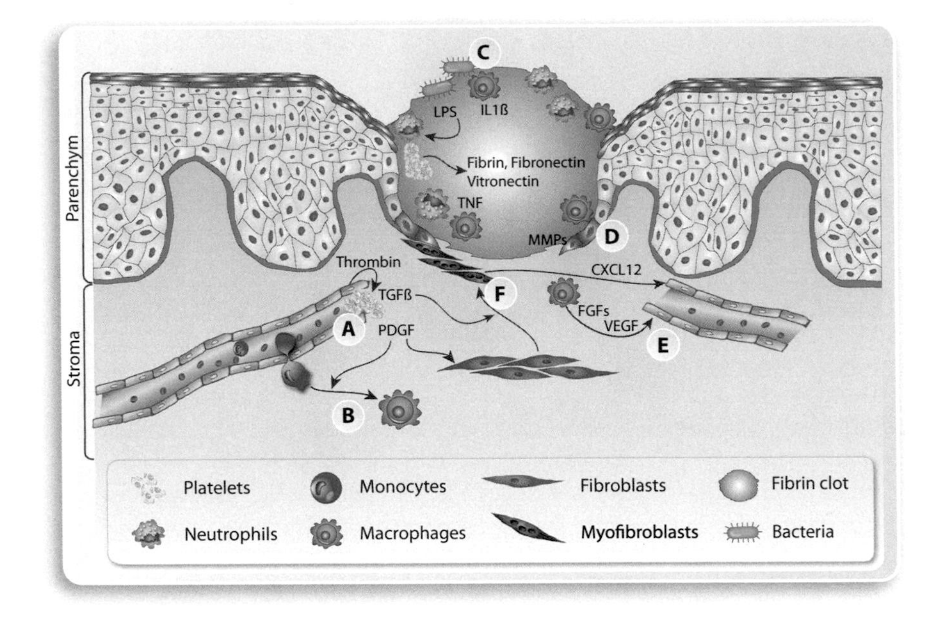

Fig. 8.1 The program of wound healing. Wound repair begins with hemostasis, where a platelet plug (**A**) prevents blood loss. Inflammation mediated by neutrophiles and macrophages (derived from monocyte differentiation (**B**)) removes debris and prevents infection (**C**). Epithelial cells undergo EMT (**D**), i.e., they start proliferation, change their morphology and migrate to close the wound gap. In addition, angiogenesis forms new blood vessels (**E**) and fibroblasts replace the initial fibrin clot with granulation tissue. Finally, the extracellular matrix is remodeled by fibroblasts and myofibroblasts causing overall wound contraction (**F**). CXCL = chemokine (C-X-C motif) ligand, FGF = fibroblast growth factor

the extracellular matrix and allow cells to leave their tissue context. Also this process is corrupted by cancer cells.

8.2 Cell Types of the Tumor Microenvironment

Organs have their own distinct ecology on the level of structure, function and biotic communities, such as microbes and malignant cells, i.e., each tissue represents a specific microenvironment not only for microbes but also for malignant cells. Accordingly, the microenvironment of tumors imposes important constraints on the development of cancer both at primary and distant sites. Importantly, malignant tumors are able to shape their own advantageous microenvironment, which protects cancer cells from external deleterious effects, such as lack of oxygen or nutrients. Thus, **the cellular composition of the tumor microenvironment represents not only**

the stage of tumorigenesis but is also important for the prognosis and possible treatment options of the cancer.

Cancer types differ in their organization, cellular cohesion and boundaries with stroma based on their cellular origin, such as epithelial or mesenchymal. Growing

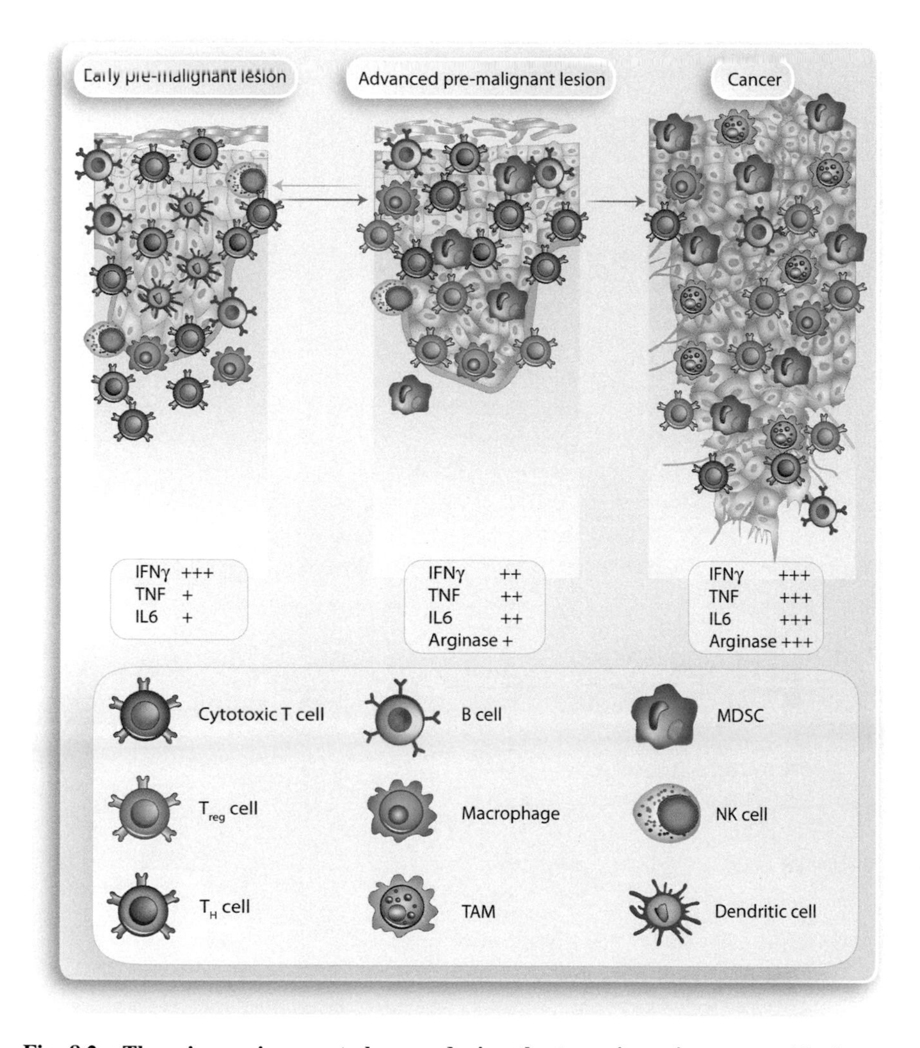

Fig. 8.2 The microenvironment changes during the tumorigenesis process. The immune microenvironment of early pre-malignant tumors is composed of cells of the innate immune system, such as NK cells, dendritic cells and macrophages, and of the adaptive immune system, such as B and T cells, while immunosuppressive T_{reg} cells and MDSCs are not often found (**left**). Advanced pre-malignant tumors contain less T and B cells but more T_{reg} cells and MDSCs (**center**). The immune microenvironment of cancer cells is dominated by T_{reg} cells and MDSCs as well as by TAMs (**right**). The change in the production of major cytokines, such as interferon γ (INF γ), TNF and IL6 is indicated below. The production of enzyme arginase is a biomarker for active MDSCs

tumors disrupt their microenvironment and recruit normal cells from surrounding tissues. Tumor-infiltrating immune cells, such as monocytes and macrophages, neutrophils, lymphocytes as well as their immature precursors, are recruited from the bone marrow to the tumor via systemic circulation. Moreover, tumors also attract tissue-resident cells, such as vascular cells like endothelial cells and pericytes (specialized mesenchymal cells that stabilize blood vessels by wrapping around endothelial cells), fibroblasts, adipocytes, nerves as well as immune cells like macrophages and mast cells. Thus, **the biology of a tumor is understood best via the integration of the function of the specialized intra- and extratumoral cell types of the microenvironment**.

The individual fitness and capacity to handle the presence of a malignant tumor critically depends on the immune system (Box 8.3). Accordingly, the intratumoral microenvironment is dominated by cells of the innate and adaptive immune system, i.e., **immune cells are a generic component of tumors**. The type and relative number of immune cells changes corresponding to the stage of the tumor within the tumorigenesis process, i.e., not only cancer cells but also cells of the microenvironment change during tumorigenesis. With progressing malignancy, the number of immune-suppressive myeloid-derived suppressor cells (MDSCs), regulatory T (T_{reg}) cells and TAMs increases, while far less cytotoxic T cells, natural killer (NK) cells and dendritic cells are found (Fig. 8.2). The change of the immune cell composition also influences the cytokine expression profile of the tumor microenvironment. This implies that **not only cancer cells themselves but also tumor infiltrating immune cells contribute to the progression and possible therapy of many cancer types** (Chap. 10).

Box 8.3: The innate and adaptive immune system. The immune system is composed of:

- biological structures, such as lymph nodes
- cell types, such as monocytes, macrophages, T and B lymphocytes (i.e., cellular immunity)
- proteins, such as complement proteins and antibodies (i.e., humoral immunity).

The immune system detects a wide variety of molecules, known as **antigens**, of potential pathogenic origin, such as on the surface of microbes, and distinguishes them from the organism's own healthy tissue. The functions of the immune system are classified into:

- the **innate immune system** is based on monocytes, macrophages, neutrophils, dendritic cells and NK cells, and uses destructive mechanisms against pathogens, such as phagocytosis anti-microbial peptides, which are supported by the complement system

- the **adaptive immune system**, which applies more sophisticated defense mechanisms, in which T and B cells use highly antigen-specific surface receptors **TCR** (T cell receptor) and **BCR** (B cell receptor), respectively, the latter finally turning into secreted antibodies. Moreover, after an initial specific response to a pathogen, the adaptive immune system creates an immunological memory that leads to an enhanced response to subsequent encounters with that same antigen.

Possible overboarding reactions of innate and adaptive immunity are counteracted by immunosuppressive MDSCs and T_{reg} cells, respectively

A malignant tumor is not only heterogenous concerning different clones of cancer cells that were created during tumorigenesis (Sect. 4.2) and the tumor infiltrating lymphocytes (TILs) but also in the cells composing the **extratumoral stroma**. Most solid malignant tumors are surrounded by connective tissue build of fibroblasts, pericytes and lymphatic and blood vessels as well as nerve cells, which are all embedded in extracellular matrix (Fig. 8.3):

- most prominent stromal constituents are cancer-associated **fibroblasts**, which derive from tissue-resident fibroblasts that are activated by TGFβ for enhanced proliferation and motility as well as extracellular matrix biosynthesis and deposition capacity. Moreover, fibroblasts secrete cytokines and chemokines of the CXCL family (Fig. 8.1). The latter has a large impact on the recruitment of immune cells to the tumor microenvironment. A minor proportion of cancer-associated fibroblasts are **myofibroblasts** (Box 8.2), which in the context of chronic inflammation contribute to pathological fibrosis
- **endothelial cells** forming tumor-associated vasculature, such as blood vessels and lymphatics (Sect. 8.3)
- **nerve fibers** are found in a variety of malignant tumors and their density is associated with poor clinical prognosis, since neuronal stimulation improves cancer cell survival
- the intestinal **microbiome** of commensal bacteria modulates malignant tumor immunity through shaping T cells, in particular T_{reg} cells. In addition to this systemic effect also the local microbiome, e.g., in the skin or the lung, plays a role
- **tissue-resident macrophages** derive from embryonic precursors that had been recruited to tissues before birth, i.e., they were present before the occurrence of a malignant tumor. In contrast, **monocyte-derived macrophages** are called up to malignant tumor (Sect. 8.4).

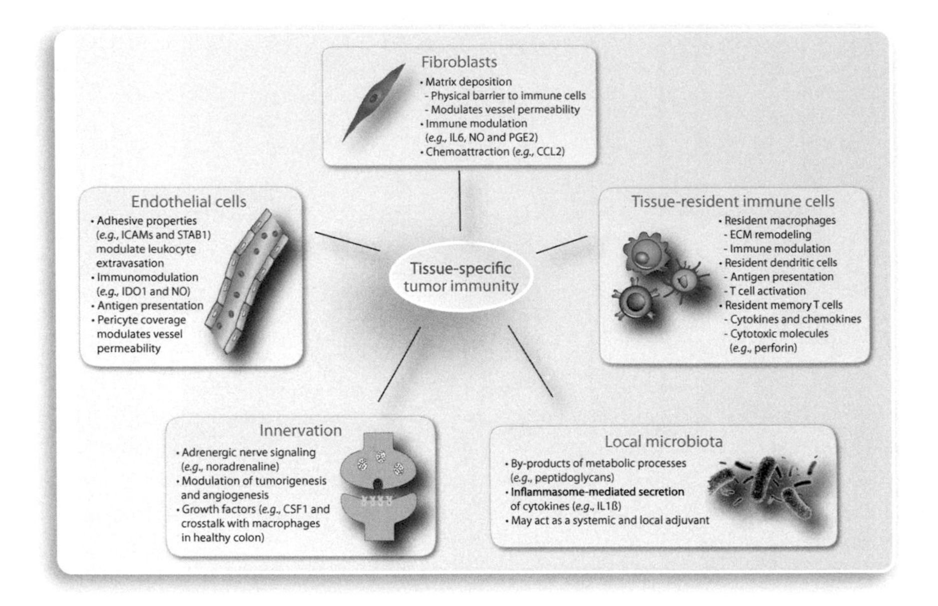

Fig. 8.3 Cellular modulators of the tumor microenvironment. The tumor immune infiltrate is controlled by blood and lymphatic vasculature of the microenvironment. Moreover, fibroblasts modulate the immune responses via the secretion of chemokines, cytokines, growth factors and ROS, as well as shaping the extracellular matrix. Tissue-resident immune cells, such as macrophages, dendritic cells and memory T cells, shape the immune responses to malignant tumors. In addition, the host's microbiome formed by commensal bacteria influences the immune responses both systemically and locally. Finally, also nerve cells affects malignant tumor cell survival, angiogenesis and the function of tumor-associated immune cells. ICAM = intercellular adhesion molecule, IDO1 = indoleamine 2,3-dioxygenase 1, NO = nitric oxide, PGE2 = prostaglandin E2, STAB1 = stabilin 1

8.3 Inducing Angiogenesis

The process of **vasculogenesis** refers to the assembly of endothelial cells into tubes and happens primarily during embryogenesis. In contrast, angiogenesis indicates the sprouting of new vessels from existing ones. In healthy adults, angiogenesis is only found transiently in the context of wound healing (Sect. 8.1) and in the female menstruation cycle. The vascular system of healthy organs shows a high specialization with clear differences in cohesiveness, pericyte coverage and expression of adhesion proteins, which is adapted to the needs of different tissues, in order to regulate vessel permeability and immune cell extravasation.

Comparable to normal tissues also malignant tumors have the need for oxygen and nutrients and have to evacuate carbon dioxide and metabolic waste. Since diffusion of these molecules is only effective over a few millimeters (Fig. 8.4, left), a tumor can grow only to a substantial size when it is vascularized (Sect. 4.1). Therefore, often hypoxia is the key angiogenic driver via the transcription factor HIF1A (hypoxia

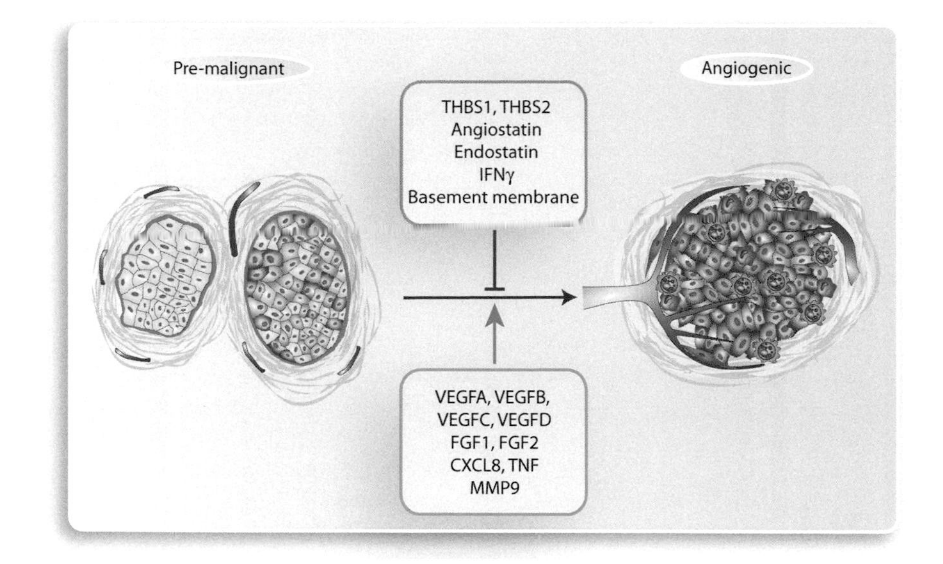

Fig. 8.4 Angiogenesis during tumorigenesis. Early-stage, pre-malignant tumors typically do not display any intratumoral vascularization, although their stroma is vascularized (**left**). In contrast, malignant tumors induce in their stroma intratumoral angiogenesis combined with immune cell infiltration, fibroblast proliferation and extracellular matrix deposition (**right**). Examples of pro-angiogenic (green) and anti-angiogenic (red) molecules are shown

inducible factor 1 subunit alpha) that stimulates the expression of VEGFA and its main receptor KDR (kinase insert domain receptor, also called VEGF receptor 2) in cancer cells as well as in cells of the microenvironment. VEGFA gradients guide mobile endothelial cells through the surrounding extracellular matrix leading to the formation of new vascular sprouts. Thus, an essential hallmark of cancer is that malignant tumor cells have to induce **permanent angiogenesis** in their surrounding microenvironment, in order to create new intratumoral blood vessels (Fig. 8.4, right). In early pre-malignant stages of tumorigenesis, such as hyperplasia and carcinoma in situ, a basement membrane (Box 8.1) separates the tumor from the vascularized tissues in the microenvironment. In these stages the cancer hallmark "inducing angiogenesis" is not reached, i.e., **blood vessels rarely infiltrate early lesions**.

Tumor progression from a small benign tumor to a large malignant stage requires an angiogenic switch (Fig. 8.4). Regulatory proteins, such as VEGFA and THBS1 (thrombospondin 1), either activate or inhibit the process of angiogenesis. These regulators are ligands to membrane receptors, such as VEGF receptors and CD (cluster of differentiation) 36, that stimulate or inhibit respective signal transduction cascades controling the growth and survival of endothelial cells. Interestingly, the VGEF signaling system comprises the receptors FTL1 (Fms related receptor tyrosine kinase 1, also called VEGF receptor 1), KDR and FTL3 (also called VEGF receptor 3) as well as their ligands VEGFA, VEGFB, VEGFC and VEGFD. The latter are encoded by oncogenes (Sect. 2.2). Other pro-angiogenic signals are members of the FGF family. Some of the pro-angiogenic signals are produced as a result of the activation of common oncogenes, such as *MYC* and *KRAS*, directly by cancer cells, whereas others origin from immune cells of the microenvironment, such as mast cells releasing VEGFA, TNF, FGF2 and CXCL8 as well as tumor-associated neutrophils producing MMP9, in order to resolve the extracellular matrix. In addition to THBS1 and THBS2, the proteins angiostatin and endostatin are endogenous inhibitors of angiogenesis.

Once angiogenesis is activated, malignant tumors differ clearly in their patterns of neovascularization, so that similar malignant tumors growing in different host organs differ substantially in their vasculature. For example, endothelial cells from distinct tissues show different expression in key surface receptors, such as VEGF receptors and TNF receptor (TNFR), and respond differently to tissue damage and inflammatory signals. Due to an unbalanced mix of pro-angiogenic signals, cancer associated blood vessels mostly show structural abnormalities, such as excessive branching, abundant and abnormal bulges and blind ends, discontinuous endothelial cell lining as well as defective basement membrane and pericyte coverage. This impaired vascular maturation leads to poor vessel functionality and incoherent tumor perfusion, i.e., **the tumor vasculature is sparse and leaky**. The tumor vasculature regulates the composition of the immune infiltrate by controling immune cell extravasation and the homing of specific immune cell subsets. Interestingly, intratumoral lymphatic vessels are mostly non-functional, since high interstitial pressure within solid malignant tumors let them collapse. However, the peripheries of malignant tumors and their adjacent normal tissues contain actively growing lymphatic vessels, which serve as channels for the seeding of metastatic cells into draining lymph nodes.

8.4 Tumor-Promoting Inflammation

Historically, tumor-associated inflammation was understood as a mechanism how the immune system tries to eradicate cancer cells. In part this still holds true (Chapter 10), but in most cases chronic inflammation associated with tumors follows the description that tumors resemble non-healing wounds (Sect. 8.1). While in normal wound healing inflammatory cells appear only transiently, chronic inflammation in malignant tumors is characterized by the persistence of TAMs and tumor-associated neutrophils (Box 8.4). These tumor-infiltrating inflammatory cells induce and sustain tumor angiogenesis and stimulate cancer cell proliferation. Thus, **inflammation is capable of pushing the development of incipient neoplasias into malignant tumors**. In addition, inflammatory cells release molecules, such as ROS, that can induce further mutations in neighboring cancer cells.

Box 8.4: Neutrophils. Neutrophils are the most abundant granulocytic cell type in human blood (40–70% of all leukocytes). They are phagocytes that belong to the innate immune system (Box 8.3). Neutrophils are the first inflammatory cells that invade a tissue that is disturbed by bacterial infection, environmental exposure or malignant tumor formation. Following a gradient of chemokines, such as CXCL8, neutrophiles migrate through blood vessels and interstitial spaces. A substantial proportion of the immune cell infiltrate in malignant tumors are tumor-associated neutrophils. Like macrophages, neutrophils produce pro-angiogenic factors and proteases. For example, neutrophiles contain granules filled with VEGFA that are rapidly released after stimulation with TNF

8.5 Deregulating Cellular Energetics

Cellular metabolism is essential for homeostasis and growth of normal cells. Metabolic pathways are supplied by abundant nutrients, such as glucose and amino acids. In the process of tumorigenesis cancer cells make a number of adaptions, in order to reprogram cellular metabolism for the most effective support of neoplastic

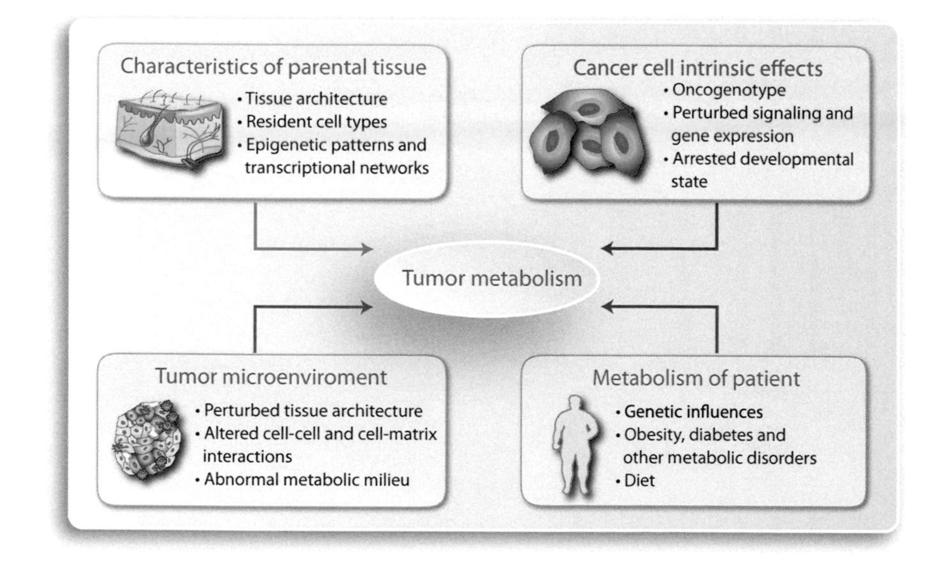

Fig. 8.5 Modulation of tumor metabolism by intrinsic and extrinsic factors. Tumor metabolism is affected by intrinsic and extrinsic factors, such as the characteristics of the parental tissue, cancer cell intrinsic effects, the tumor microenvironment and the metabolism of the patient

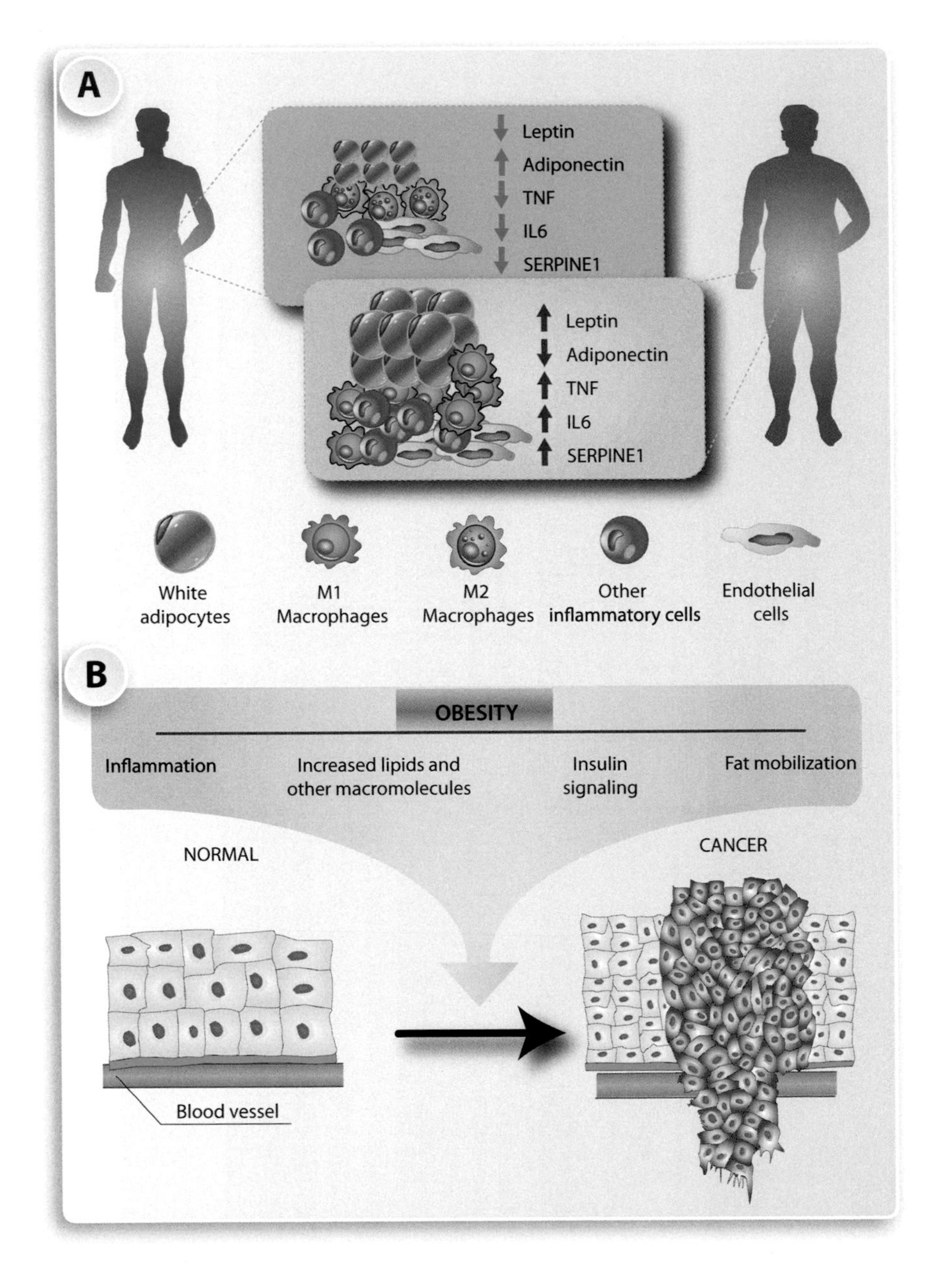

Fig. 8.6 Obesity, cancer and inflammation WAT is an endocrine and metabolic organ consisting of lipid-laden adipocytes and pre-adipocytes, macrophages, other cells of the immune system and endothelial cells (**A**). In subjects with normal weight, the adipose tissue secretes high levels of adiponectin. During weight gain, WAT expands, which mediates the infiltration of macrophages and other inflammatory cells and leads to the secretion of the cytokine TNF from macrophages. Furthermore, the secretion of IL6, SERPINE1 and leptin is also increased. Elevated inflammation, increased availability of lipids and other macromolecules, impaired insulin signaling and changes in adipokine signaling leading to fat mobilization all contribute to the conversion of epithelial cells to an invasive tumor (**B**)

proliferation. Thus, **reprogrammed metabolism implies that cancer cells improve their ability to feed themselves by using alternative mechanisms for nutrient uptake and use**. Accordingly, "deregulating cellular energetics" belongs to the extended list of 10 hallmarks of cancer. Under aerobic conditions cells use pyruvate, which was created from glycolysis of glucose, in the citric acid cycle and the oxidative phosphorylation chain for the production of some 36–38 molecules ATP per glucose molecule. In contrast, under anaerobic conditions only glycolysis is used and lactate produced, which leads to a yield of only 2 ATPs, i.e., an approximately 18-fold lower efficiency than under aerobic conditions. This deregulated glucose metabolism, referred to as "**Warburg effect**", is a metabolic switch used by cancer cells, i.e., under hypoxic conditions they consume glucose and secrete lactate generating an acidic tumor microenvironment. Accordingly, the glucose uptake by cancer cells is far higher than that of normal cells. Interestingly, some oncogenes and tumor suppressor genes encode for enzymes regulating the metabolism of nutrients, e.g., the gain-of-function mutations of the genes *IDH1* and *IDH2* can lead to the production of onco-metabolites affecting chromatin modifying enzymes (Sect. 6.1).

The interaction between all cellular compartments of the tumor microenvironment involves the exchange of metabolites that are used as source of energy but serve also for intercellular communication. For example, cancer-associated fibroblasts and adipocytes provide malignant tumor cells with nutrients, such as lipids and the amino acid alanine. Moreover, some immune cells of the tumor microenvironment get polarized by lactate, which had been secreted by glycolytic cancer cells, into an immunosuppressive phenotype (Fig. 8.5). Moreover, the overexpression of nutrient transporters can be driven by oncogenes and energy can be derived from diverse nutrient sources, such as scavenged proteins, recycled organelles and necrotic debris. Thus, **metabolic cooperativity among cancer cells or between cancer cells and stromal cells contributes to malignant tumor cell fitness**, i.e., it supports cell survival, evasion of immune surveillance and growth. Interestingly, often these effects show organ-specificity, i.e., only a few cell types respond to a given type of metabolic perturbation and get transformed.

Box 8.5: Obesity and cancer. White adipose tissue (WAT) is an important endocrine and metabolic organ consisting of both lipid-laden adipocytes and a stromal-vascular fraction, which contains pre-adipocytes, macrophages, other immune cells and endothelial cells (Fig. 8.6A). Obesity increases the size of adipocytes (hypertrophy) and number of adipocytes (hyperplasia) and is accompanied by infiltration of macrophages in the adipose tissue. Elevated levels of circulating pro-inflammatory cytokines hormones, such as IL6, TNF, leptin, resistin and serpin peptidase inhibitor, clade E (SERPINE1, also called plasminogen activator inhibitor 1), and reduced release of anti-inflammatory adipokines, such as adiponectin, are associated with obesity. Adiponectin signal transduction acts via transmembrane proteins activating the kinase AMPK (adenosine monophosphate-activated protein kinase). AMPK is a critical negative regulator of proliferation in response to energy status as it induces growth arrest and apoptosis. Obesity influences insulin signaling, which provides further energy to cancer cells, i.e., elevated insulin promotes tumor growth (Fig. 8.6B). Compared with normal fat tissue, peri-tumoral adipose tissue is highly vascularized, macrophage-rich and secretes higher levels of proteases, pro-angiogenic proteins. Moreover, elevated levels of leptin and reduced levels of adiponectin stimulate cancer proliferation

For example, the metabolic disease obesity is associated with chronic inflammation and epidemiological studies demonstrated **a 50% increase in cancer mortality in patients with a body mass index higher than 30** (Box 8.5).

Clinical conclusion: Cancer has to be understood not only on the basis of malignant tumor cells but also on its microenvironment composed of normal cells. A key component of the microenvironment are immune cells. Thus, cancer progression is majorly determined by the potency of our immune system.

Further Reading

De Palma, M., Biziato, D., & Petrova, T. V. (2017). Microenvironmental regulation of tumor angiogenesis. *Nature Reviews Cancer, 17,* 457–474.

Fane, M., & Weeraratna, A. T. (2020). How the ageing microenvironment influences tumor progression. *Nature Reviews Cancer, 20,* 89–106.

Koliaraki, V., Prados, A., Armaka, M., & Kollias, G. (2020). The mesenchymal context in inflammation, immunity and cancer. *Nature Immunology, 21,* 974–982.

Lim, B., Woodward, W. A., Wang, X., Reuben, J. M., & Ueno, N. T. (2018). Inflammatory breast cancer biology: the tumor microenvironment is key. *Nature Reviews Cancer, 18,* 485–499.

Salmon, H., Remark, R., Gnjatic, S., & Merad, M. (2019). Host tissue determinants of tumor immunity. *Nature Reviews Cancer, 19,* 215–227.

Sharonov, G. V., Serebrovskaya, E. O., Yuzhakova, D. V., Britanova, O. V., & Chudakov, D. M. (2020). B cells, plasma cells and antibody repertoires in the tumor microenvironment. *Nature Reviews Immunology, 20,* 294–307.

Vitale, I., Manic, G., Coussens, L. M., Kroemer, G., & Galluzzi, L. (2019). Macrophages and metabolism in the tumor microenvironment. *Cell Metabolism, 30,* 36–50.

Chapter 9
Metastasis and Cachexia

Abstract Malignant primary tumors become dangerous when angiogenesis allows their massive growth and in particular when some of the cancer cells spread to other organs and form metastases. Cancer cells that have obtained the hallmark "activating invasion and metastasis" use the EMT process, in order to leave the malignant primary tumor. They circulate in the blood stream until they colonize in distant organs, which often follows a specific tropism. The metastatic cascade filters for very potent few cancer cells that grow from micrometastases to macrometastases, but the clinical manifestation of metastases often takes years. Metastatic tumors take not only space and resources of their host tissues but they impact also other organs, such as skeletal muscle, via secreted cytokines and other factors. This leads in the majority of metastatic cancer patients to the muscle wasting syndrome cachexia, which finally causes multi-organ failure and death.

Keywords Metastatic cascade · Metabolic reprogramming · Metastatic niche · EMT · Metastatic tropism · Cachexia

9.1 The Metastatic Cascade

When primary malignant tumors are removed in completeness by surgery (Sect. 11.1) and no metastases have been formed, the cancer can be considered as cured. In contrast, most types of cancers are getting life-threatening when cells of a localized primary malignant tumor disseminate and form metastases at other organs. For example, the 5-year relative survival of patients with primary breast cancer is 99%, while it is only 25% with metastatic breast cancer. A primary malignant tumor may not significantly disturb the organ of origin, such as the breast, but metastases often clearly reduce the fitness of distant organs, such as bones, lung or liver. These metastatic tumors can cause neurological complications, respiratory failures, thromboses and other problems, such as cachexia (Sect. 9.4), finally leading to death.

The process of metastasis comprises a number of challenges, which are summarized as the **metastatic cascade** (Fig. 9.1):

- escape from the primary malignant tumor (intravasation) to nearby blood or lymph vessels

C. Carlberg and E. Velleuer, *Cancer Biology: How Science Works*,
https://doi.org/10.1007/978-3-030-75699-4_9

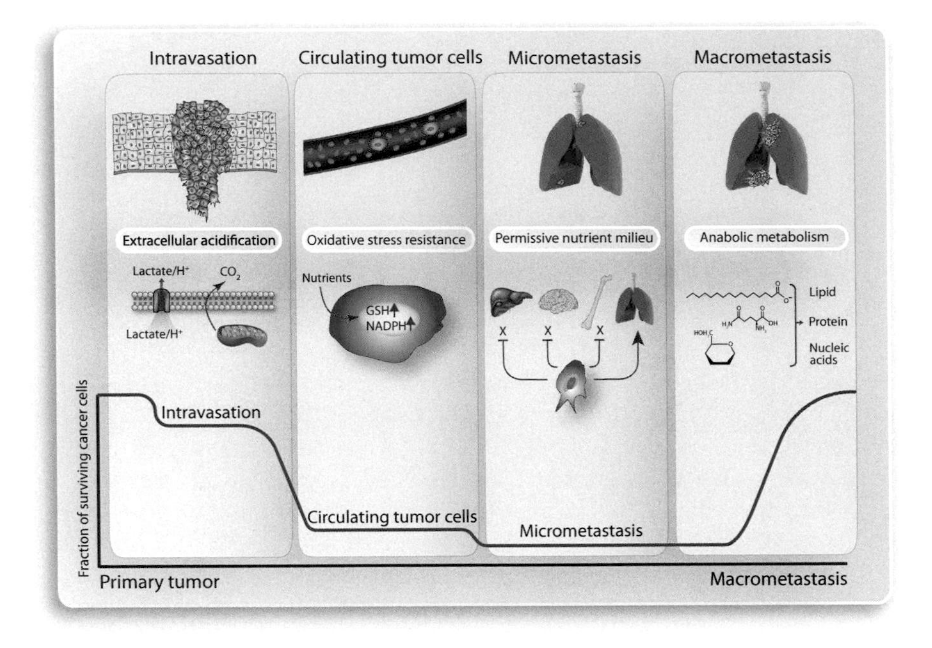

Fig. 9.1 Multiple steps of the metastatic cascade. The four main steps of the metastatic cascade (**top**) are associated with an individual survival rate of the cancer cells (**bottom**) that relates to how effective they overcome the indicated metabolic challenges (**center**)

- survival and transit through the lymphatic and hematogenous systems (circulating malignant tumor cells)
- escape of cancer cells from circulation into the parenchyma of distant tissues (extravasation)
- formation of small cancer cell nodules at distant organs (micrometastases)
- growth to large malignant tumors at distant sites (colonization with macrometastases).

These capabilities are summarized as the hallmark of cancer "activating invasion and metastasis" (Sect. 2.4), which cancer cells often acquire as last in line in the tumorigenesis process. At the site of the primary malignant tumor, cancer cells have to escape the anti-tumor immune response of the microenvironment (Sect. 10.1). Then they invade the surrounding stroma and intravasate into blood and lymph vessels, in order to circulate and spread. In the first step of the metastatic cascade, cancer cells have to degrade extracellular matrix, which is supported by extracellular acidification through the release of lactate, CO_2 and other metabolic waste products (Fig. 9.1). This decreases the number of adherens junctions between malignant tumor cells and activates proteolytic enzymes, such as MMPs, that degrade the extracellular matrix.

The intravasation step involves EMT (Sect. 9.2), which is also supported by a metabolic reprogramming of the cells, such as the activation of the enzyme UGDH (UDP-glucose 6-dehydrogenase), which reduces intracellular UDP-glucose levels, or

ASNS (asparagine synthetase (glutamine-hydrolyzing)), which converts the amino acid aspartate into asparagine. When cancer cells enter the circulation, they need to produce reducing equivalents to regenerate ROS-detoxifying metabolites. The latter are scavengers of oxidative stress, such as reduced NADPH (nicotinamide adenine dinucleotide phosphate) and the peptide GSH (glutathione), allowing the circulating cancer cells to survive the oxidizing environment of the bloodstream. The circulating cancer cells have left the assisting microenvironment of the primary malignant tumor, while in the blood or lymph as well as at distant organs they encounter a naïve, normal tissue microenvironment, which is far less supportive. The latter is often a barrier to further growth of the cancer cells. However, some new tissue microenvironments behave supportive to the circulating cancer cells and form **metastatic niches**. Thus, the choice of the metastatic sites, also referred as **metastatic tropism** (Sect. 9.3), depends on the cancer type and the specific tissue environment, such as a permissive nutrient milieu. At these sites, circulating cancer cells can extravasate and form a micrometastasis. The metastatic cells require harmony with the new microenvironment and their own needs, such as the reactivation of anabolic metabolism, in order to grow to a macrometastasis (Fig. 9.1, right). However, the success rate of this colonization is very low, which often is due to difficulties of the cancer cells assembling programs that are adaptive to the microenvironment of the distant tissue (Sect. 9.3). Thus, **the steps following cancer escape, i.e., survival, extravasation and metastatic growth, are rate-limiting for the progression of metastasis**.

In each step of the metastatic cascade cancer cells are exposed to the immune system, such as cytotoxic T cells or NK cells, which may recognize and kill them. However, metastatic cancer cells that survive the metastatic cascade have acquired strategies, such as the recruitment of immune suppressive cell like MDSCs, T_{reg} cells and regulatory B cells (Sect. 8.3). Moreover, the tumor microenvironment attracts neutrophils and macrophages (Sect. 8.4) and polarizes TAMs into type M2. The latter cells influence every step of the metastatic cascade by promoting intravasation, survival in the circulation, extravasation and growth at metastatic sites. For example, circulating cancer cells arrest in microvessels in a dormancy phase (Sect. 9.3), where they are protected from immune attack by platelets, macrophages and cytotoxic T cells. Importantly, the recruitment of immune suppressive MDSCs already to the primary malignant tumor forms a so-called **pre-metastatic niche**, which is an essential preparatory step for metastasis (Sect. 9.2). Thus, a key goal of tumor immunotherapy is the eradication of metastatic tumors, such as the use of immune checkpoint inhibitors like CTLA4 (cytotoxic T lymphocyte associated protein 4) and PDCD1 (programmed cell death 1, also called PD1), that boost the T cell response (Sect. 10.3).

9.2 Epithelial-Mesenchymal Transition

For the structural integrity, the epithelia of our body cells need to display an apical-basal polarity and are assembled in sheets laterally by tight junctions, gap junctions, desmonsomes and adherens junctions (Fig. 9.2). The latter are formed by the adhesion protein CDH1, which is encoded by a tumor suppressor gene. EMT is a reversible cellular program, in which epithelial cells are transformed into mesenchymal cells. In the healthy body, EMT is used during embryogenesis, such as gastrulation, organ morphogenesis during development, tissue regeneration and wound healing (Sect. 8.1), while in the context of cancer it enhances the metastatic potential of malignant tumor cells. EMT is induced by the expression of transcription factors of the ZEB (zinc finger E-box binding homeobox) family, SNAI (Snail family transcriptional repressor) 1 and 2 as well as TWIST1 (Twist family BHLH transcription factor 1). In various combinations these transcription factors change the expression of hundreds of genes resulting in the downregulation of proteins related to the epithelial state, such

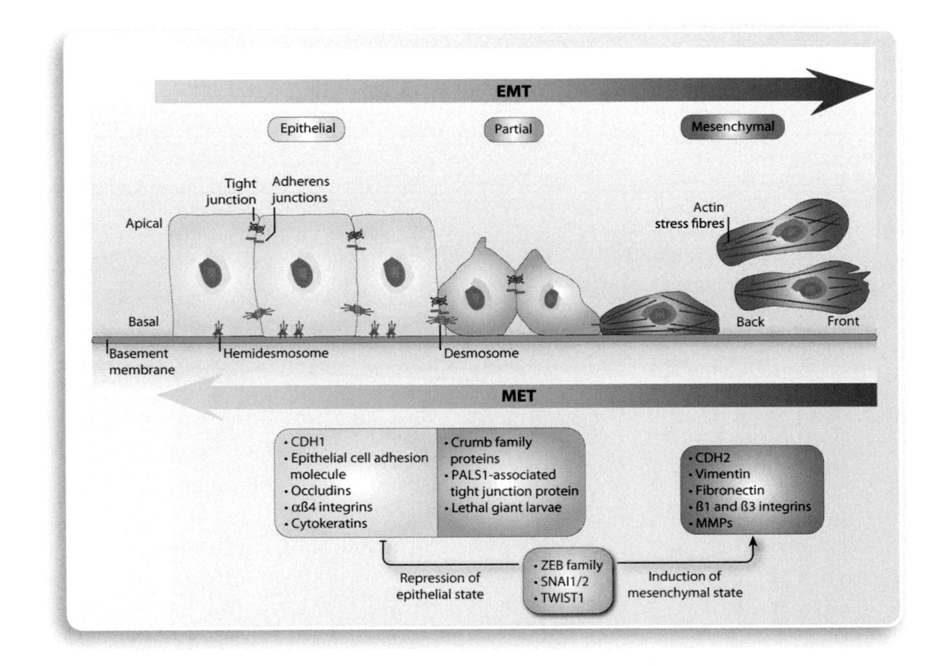

Fig. 9.2 Outline of EMT and MET. Epithelial cells are held together by tight junctions, adherens junctions and desmosomes and are connected via hemidesmosomes with the underlying basement membrane. Key expressed proteins that are essential for maintaining this status are listed below. EMT is induced by the expression of transcription factors of the ZEB family, SNAI1 and 2 and TWIST1 (**bottom**), which inhibit the expression of genes associated with the epithelial state and induce genes related to the mesenchymal state. In this way, cells become motile and acquire invasive capacities. EMT is a reversible process, since mesenchymal cells can revert to the epithelial state by undergoing MET

as CDH1, occludins and claudin, and the upregulation of proteins important for the mesenchymal state, such as CDH2 (also called N-cadherin), vimentin and fibronectin. The gene expression changes cause cellular alterations including the disassembly of epithelial cell-cell junctions and the dissolution of apical-basal cell polarity via repression of crumb family proteins, PALS1-associated tight junction protein and lethal giant larvae specifically regulating tight junction formation and apical-basal polarity. In addition, the adhesion protein CDH2, which is normally expressed during organogenesis, supports the migration of invasive cancer cells. Moreover, the expression of MMPs is increased, which degrade the basement membrane and promote cell invasion. Thus, **a transcriptional program is activated that pushes for the mesenchymal state**.

The induction of EMT disassembles epithelial cell-cell junctions and resolves cellular polarity via the repression of proteins forming tight junction, adherens junction and desmosomes (Fig. 9.2). Thus, during EMT the interactions between epithelial cells as well as their contact with the basement membrane changes, so that the cells are losing the physical contact with their environment and transform into mesenchymal cells. The progressive loss of epithelial features, such as the cobblestone appearance, leads to the acquisition of mesenchymal features, such as spindle-shaped morphology. Accordingly, mesenchymal cells extensively reorganize their actin fibers that are an important component of their cytoskeleton, so that they have stem cell-like properties, increased motility and invasive capacity, which is essential for the intravasation of cancer cells. **EMT is a step-wise process with a number of intermediary states**, but cancer cells advance only rarely to a completely mesenchymal state, i.e., mostly they reach only a quasi-mesenchymal state. Importantly, EMT is a reversible process, i.e., MET can revert mesenchymal cells back into epithelial cells. The latter is important for circulating metastatic cancer cells, which after extravasation into the distant organ turn back into their original epithelial structure. MET occurs because signal transduction cascades and epigenetic alterations lead to the repression of mesenchymal traits and the re-expression of epithelial markers, such as CDH1.

Underlying the transcriptional program that orchestrates EMT are epigenetic processes that in turn are modulated by signal transduction cascades being initiated by cytokines, chemokines and growth factors. Even in aggressive carcinoma cells EMT needs to be activated by signals from the microenvironment, such cancer-associated fibroblasts, T_H cells, cytotoxic T cells, T_{reg} cells, MDSCs and TAMs (Fig. 9.3), i.e., **without a supportive microenvironment there would be no metastasis**. This resembles in many respects the situation during inflammation and wound healing of healthy tissues (Sect. 8.1). Signaling molecules, such as cytokines TNF and IL6, the chemokine CCL18 (chemokine (C-C motif) ligand 18) and the growth factors TGFβ, VEGF, EGF and HGF (hepatocyte growth factor), act as ligands to their cognate membrane receptors expressed in cancer cells and activate respective signal transduction cascades. In this way, the signals are integrated, in order to stimulate the EMT program via the activation of the transcription factors SNAI1, ZEB1 and TWIST1. For example, TGFβ secreted by cancer-associated fibroblasts, TAMs,

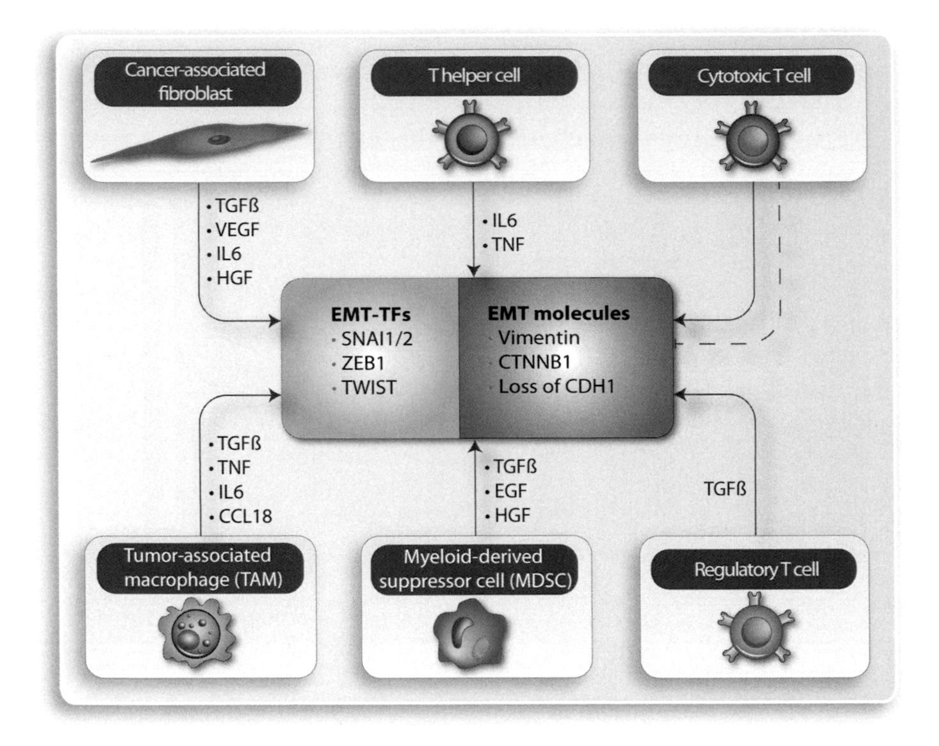

Fig. 9.3 Activation of EMT by the tumor microenvironment. The stromal cells of the tumor microenvironment, such as cancer-associated fibroblasts, T_H cells, cytotoxic T cells, T_{reg} cells, MDSCs and TAMs, secrete the indicated cytokines, chemokines and growth factors, most of which support in a paracrine fashion EMT in epithelial cells of primary malignant tumors. These signaling molecules induce EMT-stimulating transcription factors or modulate effector molecules inhibiting the epithelial state and promoting the mesenchymal state

MDSCs and T_{reg} cells activates a signal transduction cascade that results in the induction of transcription factors of the SMAD family. The latter activate mesenchymal genes, such vimentin and fibronectin, and the transcription factors SNAI1 and 2, ZEB1 and TWIST1, which in turn repress *CDH1* expression. Furthermore, TAMs secrete not only TGFβ, but also the pro-inflammatory cytokines TNF and IL6 as well as the chemokine CCL18. The latter synergize with TGFβ to promote EMT. Finally, the anti-tumor actions of cytotoxic T cells that inhibit EMT are often devastated by T_{reg} cells, M2-type TAMs and MDSCs having opposite functions.

9.3 Metastatic Colonization

Fortunately, as less as 0.01% of all cancer cells that escape from a primary malignant tumor will form macrometastases at distant organs. Factors like cellular origin, intrinsic tumor properties, circulation patterns and tissue affinities affect the metastasis process. Very critical for the efficiency of metastasis is how well circulating cancer cells form metastatic niches, i.e., colonies of micrometastases. For best adaptation of cancers cells, the metastatic niches require a modification of the extracellular matrix, a rewiring of metabolism, a polarization of stromal cells, such as TAMs, and a remodeling of blood vessels. Initially, cancer cells survive in **dormancy**, which is a cellular state of arrest or quiescence of unknown and varied duration. Finally, they re-activate growth forming macrometastases. Importantly, **metastatic colonization**, i.e., the occurrence of macrometastases, is not automatically coupled to the physical dissemination of metastatic cancer cells, since in many patients successfully spread micrometastases do not progress in their growth. This is related to the lack of adaption of the micrometastases to the microenvironment of the distant organ, which may secrete anti-growth and tumor suppressive signals.

Dormancy seems to be the default reaction of cancer cells in an unfamiliar environment like facing hypoxia in the bloodstream. The period of dormancy between the formation of a micrometastasis at a distant organ and the occurrence of a macrometastasis shows wide cancer type-specific variation ranging from months to decades. Micrometastases are smaller than a few millimeters and undetectable by clinical imaging, i.e., they are invisible. In most cases they do not cause any discomfort to the patient. Some micrometastases, such as those from the brain in the lung, may be permanently kept under control by the immune system, i.e., in an immunocompetent person they may have indefinite dormancy. In other cancers, such as that of the breast or melanoma, very long periods of latency, such as more than 10 years, are observed before life-threatening metastases occur. These long phases of dormancy are preferentially survived by self-renewing **cancer stem cells** (Box 9.1). In contrast, environmental conditions, such as inflammation caused by infections or metabolic stress during obesity, may either prevent dormancy or induce the exit from it. Thus, **not only the type of cancer but also the age and lifestyle of the patient affects the occurrence of metastases and their timing**.

Box 9.1: Cancer stem cells. A malignant tumor is formed by a number of distinct clonal subpopulations. Due to this clonal heterogeneity many tumors contain regions that differ in their degree of differentiation, proliferation, vascularity, inflammation and invasiveness. The most important cellular subclass of a primary malignant tumor is that of **cancer stem cells**. These cells are defined by their ability to seed new tumors and show widely varying abundance. Cancer stem cells have been found first in hematopoietic cancers but also occur in solid malignant tumors. In some cancers, normal adult stem cells undergo oncogenic transformation and serve as the origin of cancer stem cells, while in other cancers the stem cell character is acquired during the tumorigenesis process. For example, when epithelial carcinoma cells undergo EMT and transform into mesenchymal cells (Sect. 9.2), they obtain stem cell-like features, such as the ability to self-renew. The latter property is essential for clonal expansion of the metastatic cells at distant organs. Thus, the collection of stromal signals that induce EMT (Fig. 9.3) also stimulate the generation of cancer stem cells. Importantly, cancer stem cells show a high chemotherapeutic resistance and are often the reason for a relapse of the cancer (Sect. 11.1)

The growth rate of metastases is assumed to be comparable to their respective primary malignant tumor. Therefore, if it took 10–20 years that a primary malignant tumor obtained a significant size, for its metastases it would take about the same time. However, in many type of cancers, metastases of significant size are detected only a few years after primary malignant tumor diagnosis. This could be explained by a very early dissemination before the cancer cells acquired all hallmarks of cancer. Accordingly, there are cases where the primary malignant tumor and the metastases grow in parallel. Furthermore, there is an entity called "cancer of unknown primary", where patients are diagnosed first with metastases while the primary malignant tumor cannot be detected. Thus, **the time between diagnosis of the primary malignant tumor and its metastases depends on the time point of dissemination and the growth rate of metastases**.

Metastatic organotropism indicates that a number of primary cancers, such as in breast, colon, stomach, lung, pancreas or prostate, metastasize preferentially to particular organs (Fig. 9.4). For example, breast cancer metastases are found in bones, lung, liver and brain, while prostate cancer metastases are most often detected in bones. This tropism can be explained by the accessibility of the distant site from the position of the primary malignant tumor via the blood and lymph, i.e., the direction of the blood flow and physical vicinity are critical factors. In general, the relatively large organs liver, lung, brain and bones are primary metastatic sites of most types of cancer. Moreover, in some cases the metastatic cancer, such as pancreas cancer, cells re-seed the organ of their origin, i.e., these emigrants are coming home (Fig. 9.5).

The treatment of metastatic cancer often fails and leads to the death of the patient. However, adjuvant therapies, i.e., systemic therapies that are given in addition to the initial local therapy, such as surgery (Sect. 11.1), in order to maximize its effectiveness, sometimes showed efficiency and even in a few cases cured the patient. Nevertheless, for patients with an apparent non-metastatic cancer, relapse remains a Damocles sword. Thus, **late relapses need to be prevented, ideally by a better mechanistic understanding of the invisible phase of dormancy and growth of micrometastases**.

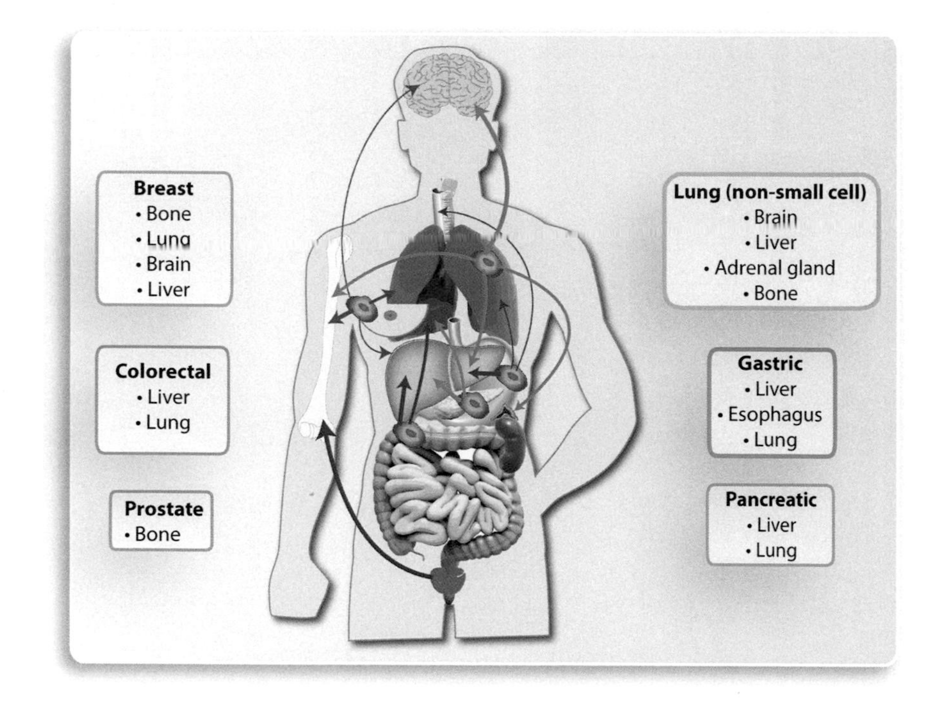

Fig. 9.4 Metastatic organotropism. Metastases originating from a particular epithelial tissue, such as breast (purple), colon (red), stomach (blue), lung (orange), pancreas (green) or prostate (brown), preferentially occur only in a limited set of the indicated distant organs. Thickness of the arrows represent the relative frequencies of metastases

9.4 Cachexia

When malignant tumors grow, they disrupt the structure and function of their surrounding tissue by secreted molecules as well as via mechanical stress. The physical traits of cancer (Box 9.2) are based on elevated interstitial fluid pressure, increased tissue volume, displacement of normal tissue and swelling of extracellular matrix. This can cause edema and increased stiffness, creates a new organization of stromal and cancer cells and their surrounding extracellular matrix adopts new organization. These physical stresses not only cause discomfort to the cancer patients but they also can affect signal transduction pathways that promote proliferation, invasiveness and metastasis of cancer cells.

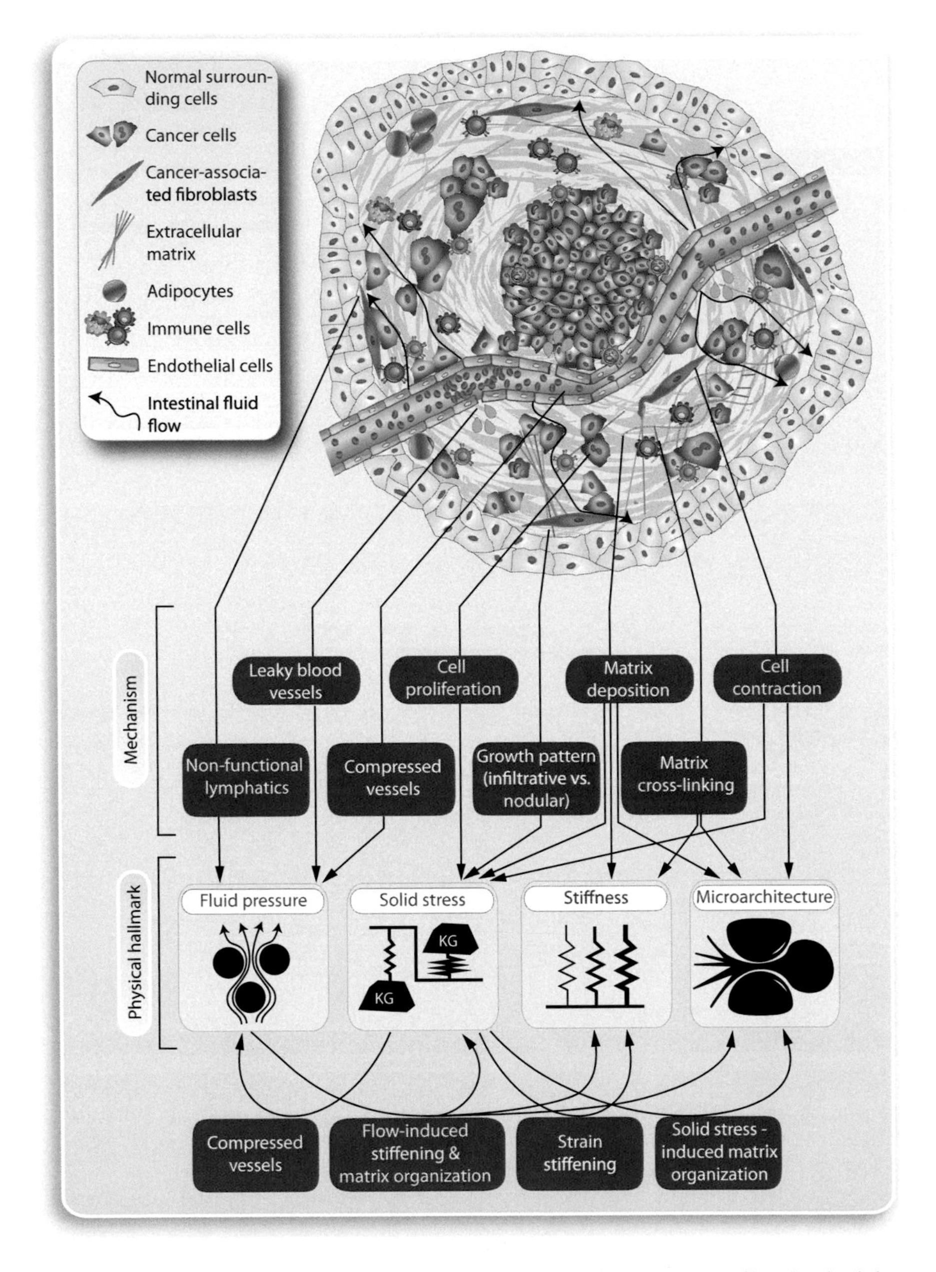

Fig. 9.5 Physical traits of cancer. Physical traits caused by malignant tumors affect the physiology of the cancer patient. More details are provided in the text

Box 9.2: Physical traits of cancer. The interactions of the cells of malignant tumor with the stroma of the tissues, in which they are embedded, cause a number of physical traits that are associated with the different types of cancer (Fig. 9.5):

- blood vessels that are abnormally leaky or compressed as well as non-functional lymphatics cause an increased pressure of interstitial fluids within the tumor and its environment
- cellular proliferation, extracellular matrix deposition, cell contraction and abnormal growth patterns increase the tumor size and result in compressive as well as tensile solid stress
- the deposition of extracellular matrix, its crosslinking and swelling due to water incorporation leads to an increased stiffness of tumors
- the architecture of the tumor bearing tissues is changed by cell contraction, extracellular matrix deposition and crosslinking.

These physical traits also interfere with each other. For example, solid stress compress blood and lymphatic vessels and contribute to increased fluid pressure in malignant tumors and increase their stiffness. In turn, fluid flow activates fibroblasts, which increase the levels of solid stress and stiffness and change the architecture of the extracellular matrix. Moreover, tensile solid stress leads to stretched and aligned extracellular matrix

Since complete cancer elimination is rarely achieved in metastatic cancer patients, more than 80% of them, in particular those suffering from pancreatic, gastrointestinal, colon and lung cancer, develop **cachexia**. This progressive loss of muscle mass, referred to as muscle atrophy, and function often also includes the loss of adipose tissue mass. **Cachexia depends on signals derived from metastatic malignant tumors,** i.e., **its onset and maintenance requires the presence of a substantial tumor mass in the patient**. Malignant tumors secrete various signals, such as growth factors, cytokines, hormones, exosomes and metabolites, that are able to reprogram the physiology, metabolism and immune response of distant organs. This includes skeletal muscle or adipose tissue that are only seldomly the target of metastases. Some of these signaling molecules, such as the pro-inflammatory cytokines TNF (via NF-κB) and IL6 (via STAT3), the anti-inflammatory cytokine TGFβ (via SMAD proteins) and extracellular vesicles (EVs) transporting HSP (heat shock protein) 70 and HSP90 that activate the MAPK pathway, directly promote muscle atrophy (Fig. 9.6, left). In addition, there are indirect effects on muscle wasting via the peptide hormone PTHrP (parathyroid hormone-related protein) and IL6 stimulating the browning of white adipose tissue expending energy via thermogenesis and through the TGFβ superfamily members GDF (growth differentiation factor) 11 and GDF15 promoting anorexia via the regulation of appetite in the hypothalamus (Fig. 9.6, right). Moreover, insulin resistance of skeletal muscle and metabolic reprogramming of the liver via increased production of inflammation-inducing acute phase response proteins,

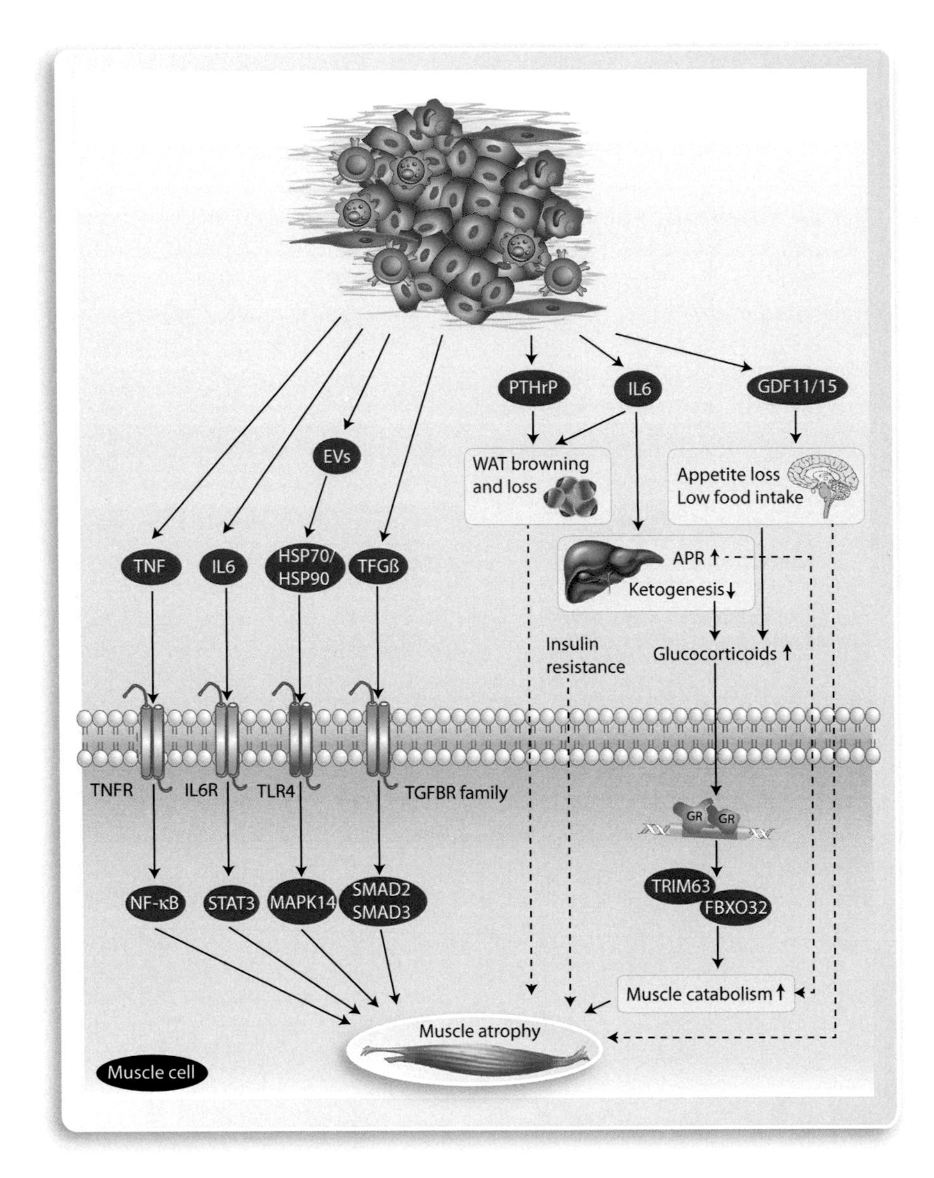

Fig. 9.6 Mechanisms of muscle atrophy in cancer. Signals produced by cells of a malignant tumor and its microenvironment, such as cytokines, hormones, growth factors and extracellular vesicles, affect distant organs, such as liver, brain and adipose tissue, and contribute to muscle atrophy. This activates pro-cachectic programs in muscles that either directly via interaction with muscle cells stimulating catabolic pathways or suppressing protein synthesis (**left**) or indirectly via the metabolic reprogramming of secondary organs (**right**) induce muscle wasting. APR = acute protein response, IL6R = IL6 receptor, TGFBR = TGFβ receptor, TLR4 = Toll-like receptor 4

reduced ketone body formation and increased glucocorticoid levels also commit to muscle atrophy. Glucocorticoids act via the transcription factor GR (glucocorticoid receptor), target genes of which, such as *TRIM63* and *FBXO32* (F-box protein 32), mediate muscle wasting. Thus, in response to both direct and indirect signals the protein homeostasis of skeletal muscle cells is in misbalance, which results in increased protein degradation via the ubiquitin-proteasome and autophagy pathways and decreased protein synthesis.

Skeletal muscle is the main organ affected by the wasting process during cachexia, but in addition also white and brown adipose tissue, liver, brain, gut and heart are concerned (Fig. 9.7). This leads, e.g., to abnormalities in the function of the heart, changes in the synthesis of liver proteins and alterations in the levels of peptide hormones affecting the hypothalamus. Accordingly, the brain-based control of appetite, satiation, taste and smell, which affect food intake and contribute to the altered energy balance of the cancer patient results in anorexia. Moreover, increased inflammation in many organs activates anorexigenic pathways and decreases orexigenic routes. Thus, **cachexia is a multi-organ syndrome**.

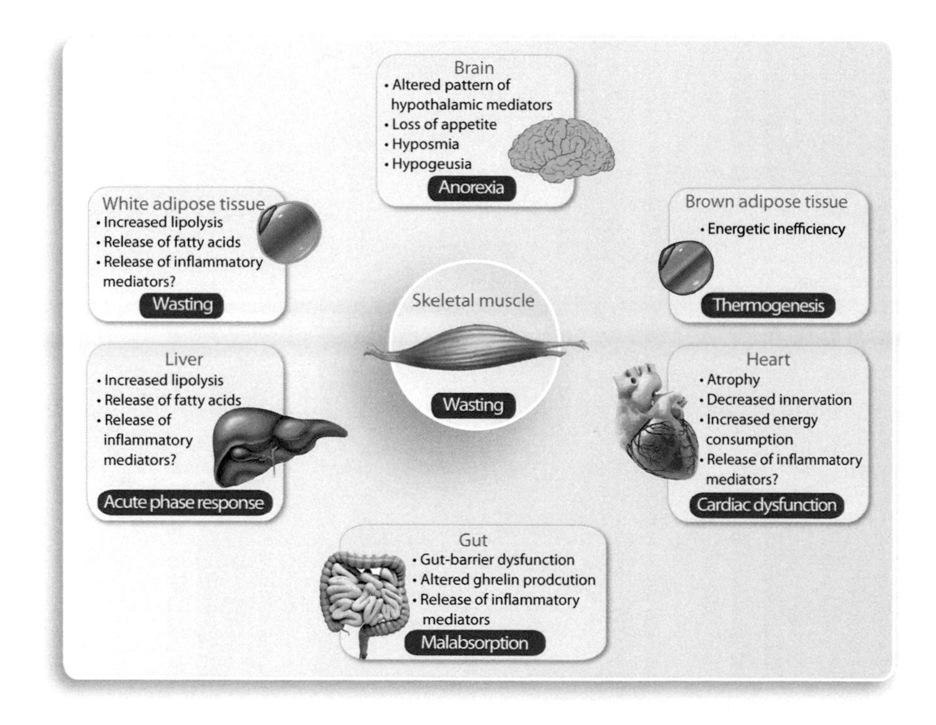

Fig. 9.7 **Cachexia is a multi-organ syndrome.** The wasting of skeletal muscle and adipose tissue in cachexia affects other organs, such as liver, gut, heart, brown adipose tissue and brain via the indicated dysfunctions

The function of muscles is vital for breathing, movement, chewing and swallowing food. Therefore, muscle wasting after the development of **cachexia leads to a significantly diminished quality of life of the cancer patient**. Moreover, malfunctional muscles lower the tolerance of the patients to therapy (Sect. 11.1). Finally, weakened heart and lung muscles often lead to deaths from cardiac and respiratory failures. Thus, **cachexia associated with metastatic cancers has devitalizing and lethal consequences for the patient**.

Clinical conclusion: Most types of cancer get life-threatening only when they form metastases. When a primary tumor is diagnosed there is the risk that it had already been spread to other organs. Thus, a cancer should not be treated and understood exclusively as a local disease but always has to be viewed systemically. This is of importance for the elimination of potential metastases.

Further Reading

Biswas, A. K., & Acharyya, S. (2020). Understanding cachexia in the context of metastatic progression. *Nature Reviews Cancer, 20,* 274–284.

Dongre, A., & Weinberg, R. A. (2018). New insights into the mechanisms of epithelial–mesenchymal transition and implications for cancer. *Nature Reviews Molecular Cell Biology, 20,* 69–84.

Faubert, B., Solmonson, A., & DeBerardinis, R. J. (2020). Metabolic reprogramming and cancer progression. *Science, 368*, aaw5473.

Kai, F., Drain, A. P., & Weaver, V. M. (2019). The extracellular matrix modulates the metastatic journey. *Developmental Cell, 49,* 332–346.

Kitamura, T., Qian, B.-Z., & Pollard, J. W. (2015). Immune cell promotion of metastasis. *Nature Reviews Immunology, 15,* 73–86.

Klein, C. A. (2020). Cancer progression and the invisible phase of metastatic colonization. *Nature Reviews Cancer, 20,* 681–694.

Lu, W., & Kang, Y. (2019). Epithelial-mesenchymal plasticity in cancer progression and metastasis. *Developmental Cell, 49,* 361–374.

Massague, J., & Obenauf, A. C. (2016). Metastatic colonization by circulating tumor cells. *Nature, 529,* 298–306.

Nia, H. T., Munn, L. L., & Jain, R. K. (2020). Physical traits of cancer. *Science, 370*, eaaz0868.

Turajlic, S., Sottoriva, A., Graham, T., & Swanton, C. (2019). Resolving genetic heterogeneity in cancer. *Nature Reviews Genetics, 20,* 404–416.

Chapter 10
Cancer Immunity

Abstract A potent immune system is not only protecting us from infectious diseases but also performs daily surveillance of our body for transformed cancer cells. The core mechanism of this cancer immunity is the recognition of neoantigens on the surface of cancer cells and the elimination of these cells by cytotoxic T cells. However, immune checkpoints are often modulated by negative regulators, such as CTLA and PD1, which prevent an efficient action of the cytotoxic T cells. Monoclonal therapeutic antibodies that block CTLA and PD1 enhance the anti-tumor response of T cells and show remarkable clinical effects. The most recent therapeutics are T cells of the patient that have been engineered with a chimeric antigen receptor (CAR) and are applied preferentially in hematological malignancies. Further advances in the understanding of immunology and genetic engineering technologies will lead to even more sophisticated immune therapies against cancer with less systemic side effects than classical chemotherapies.

Keywords Immune system · Cytotoxic T cells · Neoantigens · Immune checkpoint · Monoclonal antibodies · CTLA · PD1 · Cellular immune therapies · CAR T cells

10.1 Outline of Cancer Immunity

Animals are equipped with an immune system (Box 8.3) in order to effectively fight against infections by microbes, such as viruses, bacteria, fungi and parasites. The immune system is a system of biological structures, such as the lymphatic system, specific cell types, such as leukocytes (cellular immunity), and proteins, such as antibodies and complement proteins (humoral immunity). Microbes mostly have a significantly higher growth rate than cells of the host, often secrete or present exo- and endotoxins and cause severe illness to the host or even kill him. Due to the immediate danger, the response of immune cells to a direct or indirect contact with microbes and other molecules with an antigenic pattern has to be stronger than in most other physiologic situations. **Without an effective immune system, we would very likely die from infectious diseases before we reach maturity, i.e., we would not get old enough getting children and raising them**.

C. Carlberg and E. Velleuer, *Cancer Biology: How Science Works*,
https://doi.org/10.1007/978-3-030-75699-4_10

In a number of diseases, such as bacterial sepsis, the overreaction of the immune system makes us severely ill or may even lead to death. This overreaction can also be observed after an infection with the virus SARS-CoV 2 (severe acute respiratory syndrome coronavirus 2). The acute overreaction of the immune system in the adult patients is known as COVID-19 (coronavirus disease 2019) and its more latent variant MIS-C (multisystem inflammatory syndrome in children), which is prevalent in children. These examples show that **our immune system has the potential to kill us**. Other non-desired malfunctions of the immune system are allergy, autoimmune diseases and graft rejections in case of organ transplantations (Table 10.1). In fact, the molecular mechanisms that allow the immune system to recognize cells from a genetically different person and eliminate them, is a special case of the constant surveillance of our body for the appearance of transformed cancer cells and their destruction. Thus, the detection and elimination of abnormal cells by the immune system is part of its normal function. This is supported by the observation that patients who are

Table 10.1 Benefits and disadvantages of a potent immune system. The immune system of each individual has its own efficiency to respond to challenges by microbes, allergens, organ transplantations and emerging cancer cells

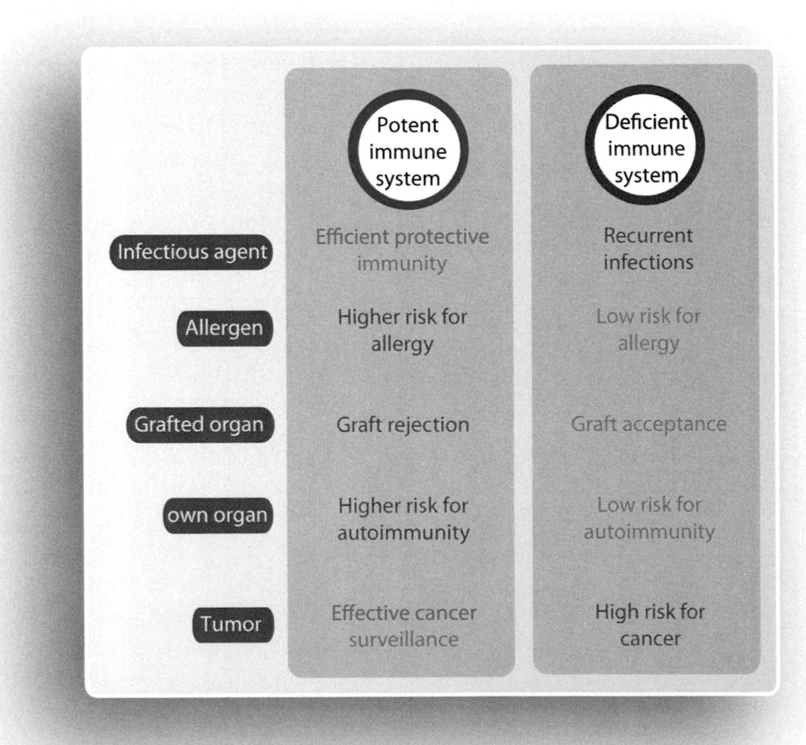

	Potent immune system	Deficient immune system
Infectious agent	Efficient protective immunity	Recurrent infections
Allergen	Higher risk for allergy	Low risk for allergy
Grafted organ	Graft rejection	Graft acceptance
own organ	Higher risk for autoimmunity	Low risk for autoimmunity
Tumor	Effective cancer surveillance	High risk for cancer

prescribed with immunosuppressive drugs, e.g., after an organ transplantation, have a significantly higher risk of getting a number of different cancers (Box 10.1). In particular, when the development or function of cytotoxic T cells, T_H1 helper cells or NK cells is compromised, the cancer rate clearly increases. For example, HIV-positive individuals that are severely immunocompromised by the outbreak of AIDS (acquired immunodeficiency syndrome) often develop Kaposi sarcoma. Importantly, immune cells are a regular component of the microenvironment of malignant tumors (Sect. 8.2). Accordingly, the occurrence of TILs are a sign that the immune system is responding to the presence of the malignant tumor. Thus, **a potent immune system should protect us throughout our life from cancer**.

Box 10.1: Cancer incidence of immunosuppressed patients. Patients receiving an organ transplant need to suppress their immune system by long-term therapy, e.g., with the calcineurin inhibitor cyclosporin or comparable drugs, in order to avoid rejection of the organ, referred to as host-versus-graft disease. Moreover, some infectious diseases destroy the immune system, such as the drastic reduction of T_H cells by HIV. In both cases the immune system is less able to detect and fight pre-malignant and cancer cells. This increases the risk for many types of cancer. Moreover, immunosuppressed patients are prone to infectious diseases and existing infections can become chronic. Accordingly, not only the direct effect of the suppression of the immune system increases the cancer risk but also the indirect effect via chronic inflammation plays an important role in the tumorigenesis in those patients. Most of the cancers arising in these immunosuppressed patients are non-Hodgkin lymphomas and cancers in the lung, kidney and liver, i.e., they represent rather non-common cancer types. Patients with a hematopoietic stem cell transplant, who developed graft-versus-host disease also have an elevated cancer risk due to the misbalance of their immune system. Thus, **the immune system has an important role in the fight and prevention of cancer**

The huge count of some 70,000 daily lesions to our genome per cell (Sect. 4.3) and approximately 3000 billion cells forming our body makes it very likely that every day transformed cells are established, even if potent DNA repair systems are constantly active. Thus, **without an effective immune system each of us would suffer in early life from different types of cancer**.

The interaction of the immune system with cells of a malignant tumor is understood as a series of events summarized as the cancer-immune cycle (Fig. 10.1). During aging and in particular during tumorigenesis, cells are accumulating thousands of passenger mutations and a few driver mutations. In a typical solid malignant tumor, some 30-70 of these mutations affect the coding region of proteins and can lead to new substructures in proteins. **These structures are referred to as neoantigens, since they are foreign to the immune system and have the potential inducing**

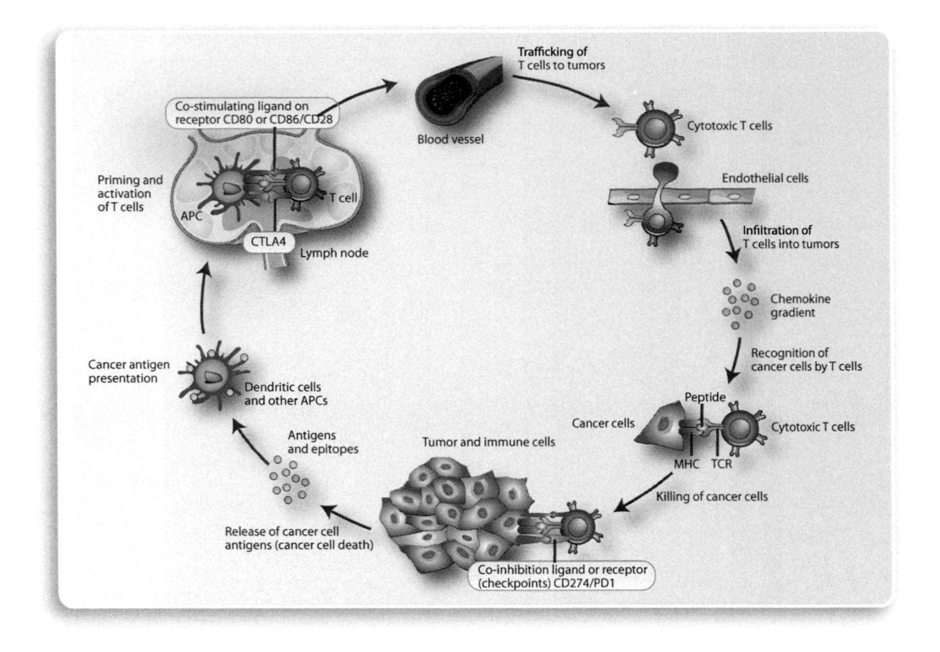

Fig. 10.1 The cancer-immune cycle. The interaction of the immune system with cancer cells involves a number of consecutive steps: (i) cancer cells (**bottom**) present or release neoantigens that are detected and ingested by dendritic cells, (ii) in lymph nodes dendritic cells present on their MHC complexes neoantigen peptides to T cells (**left**), (iii) T cells carrying a TCR fitting to the MHC-peptide complex undergo clonal expansion and circulate then in the bloodstream until they are attracted by a chemokine gradient targeting them to the cancer cells (**top**) and (iv) T cells specifically recognize cancer cells and kill them (**right**). Key molecules and cells are indicated, more details are provided in the text. APC = antigen presenting cell

its strong response. Accordingly, T cells that are specific for neoantigens are not negatively selected in the thymus, since they are not known from normal cells (self).

In the regular turnover of proteins, neoantigens are presented as peptides bound to MHC (major histocompatibility complex) proteins on the surface of cancer cells or are secreted as a whole protein when the cell is dying. Secreted neoantigens are taken up via phagocytosis by tissue-resident dendritic cells, which then move via the lymph system to the closest lymph node (referred to as sentinel lymph node). Lymph nodes contain a large number of T cells that are all different clones, i.e., they vary in the exact structure of the antigen recognition region of their TCR. Dendritic cells present the neoantigen peptides via MHC proteins to a large number of T cells until one is found, the TCR of which recognizes with high affinity the MHC-neoantigen peptide complex (Sect. 10.2). This so-called **immune checkpoint** is supported positively and negatively via the contact of pairs of auxiliary membrane proteins, such as a member of the B7 family of co-stimulatory proteins like CD80 and CD86 on dendritic cells as well as CD28 (activating) and CTLA (inhibiting) on T cells (Sect. 10.3). The purpose of negative regulators of this immune checkpoint is to avoid possible overreactions of the immune system, such as observed in allergy and autoimmune disease. Activated T cells undergo clonal expansion and thousands to millions of identical cells go into circulation via the bloodstream until a chemokine gradient is directing them to cancer cells. The specific recognition of cancer cells is again facilitated by the immune checkpoint formed by TCR and MHC proteins presenting the neoantigen peptide, by which the T cell clone had been selected. Activating and inhibitory auxiliary receptors, such as CD274 (also called PDL1) on the cancer cells and PD1 on the T cells, modulate the efficiency of the T cell-cancer cell contact. In case of an activating contact, the cytotoxic T cells secrete granulysin, perforin 1 and serine proteases of the granzyme family, which **induce apoptosis within the cancer cells**.

In addition to the specific recruitment of anti-tumor cytotoxic T cells, the microenvironment contains a number of cells of the innate and adaptive immune system. As already discussed in Sect. 8.2, the immune microenvironment of early pre-malignant tumors clearly differs from that of aggressive carcinomas (Fig. 8.2). Main differences are a shift from the dominance of anti-tumor immune cells, such as cytotoxic T cells, NK cells, dendritic cells, neutrophils and pro-inflammatory M1-type TAMs to a governance by tumor supportive M2-type TAMs, immunosuppressive T_{reg} cells and MDSCs (Fig. 10.2). The change in the immune microenvironment is also visible on the level of signaling molecules that are secreted by aggressive tumors, such as the cytokines INFγ, TNF and IL6 as well as the chemokines CXCL1, 5, 9, 10, 11 and 13. Cytokines create a pro-inflammatory milieu of the malignant tumor that also affects neighboring organs (Sect. 9.4), while chemokines guide further immune cells to the malignant tumor. IL21 and CXCL13 levels are used most often as biomarkers (Box 10.2) of a good **immune contexture** predicting the patients' survival. Accordingly, the immune contexture provides an advance to the classical tumor classification system (Box 4.2). Importantly, malignant tumors with a high mutational rate, such as melanoma and NSCLC, have a large load with neoantigens and recruit far more immune cells than cancers with lower numbers of mutations. Thus, **the tumor immune infiltrate is an important characteristic for the respective cancer type as well as for its prognosis and possible therapy**.

Box 10.2: Immunologic biomarkers of cancer. In general, a biomarker is a substance, structure or process that can be measured in the body or its products and influences or predicts the incidence of outcome or disease. In cancer, biomarkers are primarily used for **prognosis**, i.e., they are factors that influence positively or negatively the clinical outcome of the patients. For example, non-proliferating versus proliferating tumor-associated cytotoxic T cells have a negative or positive prognostic value, respectively. Moreover, T_H1 cells and their main cytokine INFγ predict a good clinical outcome in all cancer types. In addition, the presence of tumor-associated NK cells predicts an increased survival in patients with colorectal and prostate cancer, while increased numbers of tumor-associated neutrophils provide a negative prognosis. Thus, **the immune contexture contains a number of useful prognostic biomarkers**. However, in most cases, these biomarkers are not predictive for therapeutic effectiveness, where other biomarkers need to be applied

Taken together, immunity is a rapidly responding system that is capable to remove not only a load of microbes but also kills huge numbers of cancer cells generated in our body. **This process of cancer immunosurveillance works perfectly for every second of us and provides life-long absence of any detectable malignant tumor**.

In the same way as pathogenic microbes find mechanisms to escape from immunosurveillance and elimination, also cancer cells often adapt the hallmark "avoiding immune destruction". Examples for the latter are:

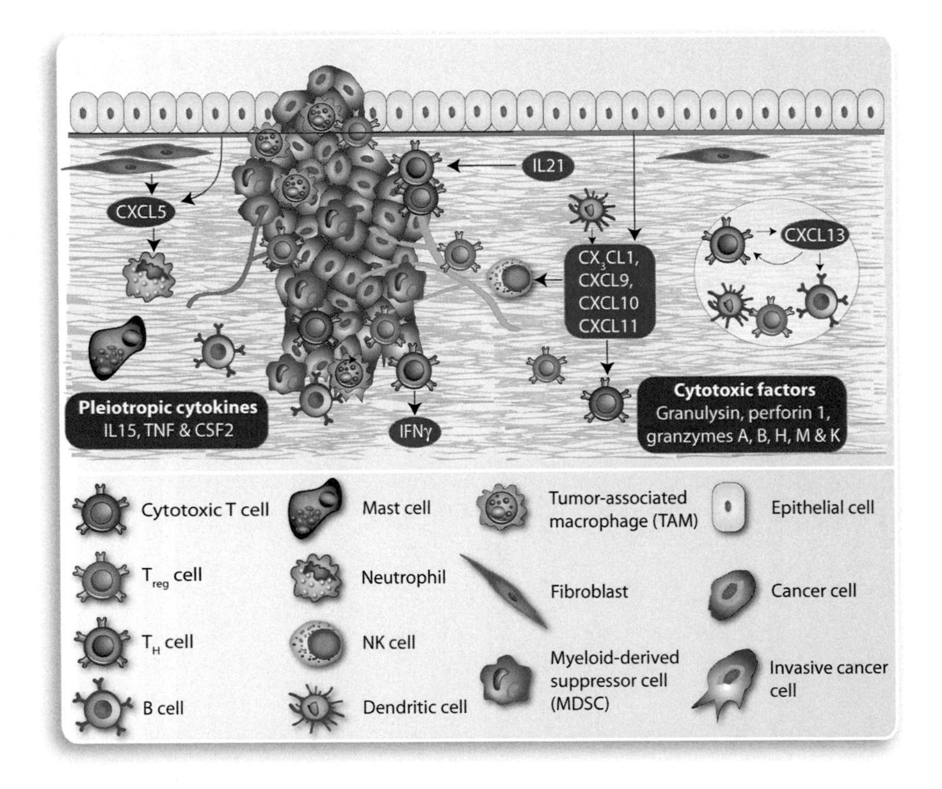

Fig. 10.2 Main immune cells shaping cancer progression. Immune cell types are shown belonging to the microenvironment of malignant tumors. Cytokines and chemokines with a positive prognosis on cancer are in blue boxes, while those with negative impact are in red boxes. Further details are provided in the text

- genetic changes that make cancer cells less visible to the immune system, such as the downregulation of MHC complexes carrying neoantigens
- upregulating the expression of membrane proteins that inhibit the immune checkpoint, such as CD274
- induce a change in the pattern of TILs from anti-tumor cytotoxic T cells to pro-tumor T_{reg} cells.

However, even in cases when immunosurveillance has failed in removing a malignant tumor in an early stage, immunotherapies, such as:

- boosting the recognition of cancer neoantigens and creating tumor vaccines (Sect. 10.2)
- blocking immune checkpoint inhibitors by monoclonal antibodies (Sect. 10.3)
- the use of engineered CAR T cells (Sect. 10.4)

belong to the most promising anti-cancer treatments. Therefore, **immunotherapy may be the most effective strategy for the elimination of metastatic cells in completeness**.

10.2 Recognition of Tumor Antigens

Healthy adults may have up to 10 billion different clones of naïve T cells (Box 10.3) in their lymphoid organs and in circulation. As a whole these T cells are able to recognize about this number of variant protein antigen structures, i.e., our immune system is well prepared to react on a huge number of putative antigens. These antigens may be of environmental origin, such as proteins on the surface of microbes that we got infected with, but they may also be created by our own cells. The regular turnover of all intracellular proteins via the ubiquitin-proteosome system in the cytosol results in a large number of smaller peptides, some of which are loaded in the endoplasmatic reticulum to MHC-I complexes and finally are exposed on the cell surface. In this way, fragments of intracellular proteins are visible to the immune system and define self-antigens. Central tolerance mechanisms taking place in the thymus prevent that any naïve T cell gets into circulation, the TCR of which recognize by chance these self-peptides. Accordingly, **normal cells should not cause any response of cytotoxic T cells** (Fig. 10.3A).

Box 10.3: T and B cell clones. The adaptive immune system is composed of T and B cells that use clones of highly antigen-specific receptors, TCRs and BCRs, for the recognition of their molecular targets. This implies that each naïve T and B cell carries one version of an antigen receptor that is unique for a putative antigen. Importantly, T and B cells "learn" during their schooling phase in the thymus or the bone marrow, respectively, not to recognize via their expressed TCRs and BCRs any antigen presented by normal body cells, referred to as "self". T and B cells that by chance recognize self-proteins are removed by this central tolerance mechanism through the induction of apoptosis. Naïve B cells are able to recognize via their BCR a larger number of different molecule classes, such as whole proteins, carbohydrates and DNA. Naïve T cells bind via their TCR only peptides of a size in the range of 8-30 amino acids that are presented on a MHC protein on the surface of a target cell. There is a constantly new production of naïve T and B cells, which die within a few weeks, if they are not happen to get presented a specific antigen. However, at any moment of time our immune system is composed by approximately 10 billion different T and B cell clones that are well prepared to respond to a respective high number of putative antigen structures. When a naïve T or B cell gets in contact with an antigen, to which their TCR or BCR has high affinity, they start to proliferate and differentiate into effector cells. In this way, clones of T or B effector cells are generated that are comprised of thousands to millions of cells carrying identical copies of highly specific antigen receptors

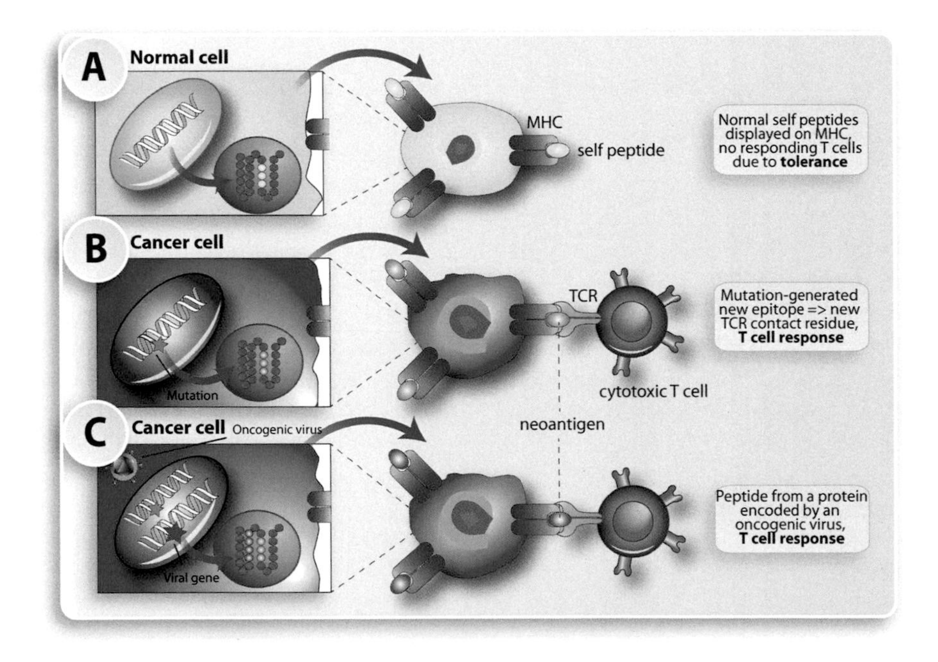

Fig. 10.3 Occurrence of neoantigens in cancer cells. T cells have learned via central tolerance mechanisms during their schooling phase in the thymus not to react to peptides being presented on MHC proteins, which originate from regular cellular proteins (self) in normal cells (**A**). In cancer cells, neoantigens can be produced by non-synonymous driver or passenger mutations in the coding region of regular proteins, which cause an amino acid exchange (**B**), or by new proteins, the genes of which are introduced by oncoviruses (**C**). In the two latter cases MHC-presented peptides cause a response of the TCR of cytotoxic T cells

In contrast, when somatic cells accumulate driver and passenger mutations, some of these affect in a non-synonymous way the coding region of proteins and cause amino acid exchanges. In addition, some of these mutations may create frameshifts, splice variants, gene fusions and other processes that result in changes of endogenous proteins. When the respective proteins are degraded, new types of peptides are formed and presented by MHC proteins on the cell surface (Fig. 10.3B). These neoantigens are exclusively recognized by those clones of cytotoxic T cells that fit with their TCR to the MHC-neoantigen complex. The specific contact of cytotoxic T cells starts an intracellular signal transduction cascade that results in the release of apoptosis-inducing granzymes, i.e., **the neoantigen carrying cancer cell is eliminated**. Similarly, when cells are transformed by an oncovirus they produce large amounts of viral proteins, some of which also get degraded, so that neoantigens can be loaded on MHC proteins (Fig. 10.3C). The elimination of the oncovirus-infected cells follows the same principles as described for neoantigens derived from mutated proteins.

The elimination of oncovirus-transformed cells by cytotoxic T cells can be considered as a special case of immunosurveillance for cells infected by regular viruses, i.e., **for the immune system there is no mechanistic difference in its fight against transformed cancer cells or virus infections**. Since vaccination is an effective therapy against virus infections, there may be also the chance for a vaccination against cancer. Such an anti-cancer vaccination comprises not only a cancer preventive therapy against tumor viruses, such as EBV (Epstein-Barr virus), HBV and HPV (Sect. 1.5), but also the application of vaccines based on tumor neoantigens. Malignant tumors with a larger number of mutations, such as melanoma, carry many neoantigens. These cancers are more efficiently recognized by the immune system and are the best targets of cancer treating and protecting vaccines. Interestingly, the characterization of neoantigens of malignant tumors allows an optimized use of immune checkpoint inhibitors (Sect. 10.3) and improves other forms of immunotherapy, such as CAR T cells (Sect. 10.4).

On the basis of results obtained with next-generation sequencing methods (Box 5.2), such as RNA-seq, exome sequencing and whole genome sequencing, a cancer patients' mutational burden can be determined and neoantigens that are specific for the respective malignant tumor can be envisioned. Moreover, these data also allow the determination of variants of *HLA* (human leukocyte antigen) genes encoding for MHC proteins and can predict their neoantigen peptide affinity. **This molecular profiling information allows the design of precise cancer treating vaccines on the basis of DNA or RNA, peptides or dendritic cells**. Accordingly, the information about tumor-specific neoantigens can be used in a number of different approaches, such as

- dendritic cells presenting neoantigen peptides
- viruses encoding neoantigen peptides on their surfaces
- vaccines containing neoantigen peptides and antibodies
- T cells with reactivity directed against neoantigen peptides.

However, targeted therapies are very costly and in case of metastatic cancer they may not be sufficient, i.e., they add to the substantial costs of other therapies. Therefore, **also from the economic point of view it would be more efficient to boost the whole immune system.**

10.3 Monoclonal Antibodies in Cancer Immunotherapy

In contrast to cytotoxic T cells that physically interact via their TCR with neoantigen-loaden MHC proteins on the surface of cancer cells (Sect. 10.2), after initial antigen presentation B cell clones follow the strategy to differentiate into plasma cells. In this differentiated stage they produce huge amounts of antibodies that have the same antigen specificity as their membrane bound BCR, i.e., antibodies are secreted forms of BCRs. The Y-shaped structure of antibodies (Fig. 10.4A) with two identical antigen binding sites provides them with a remarkably high affinity for their specific antigen

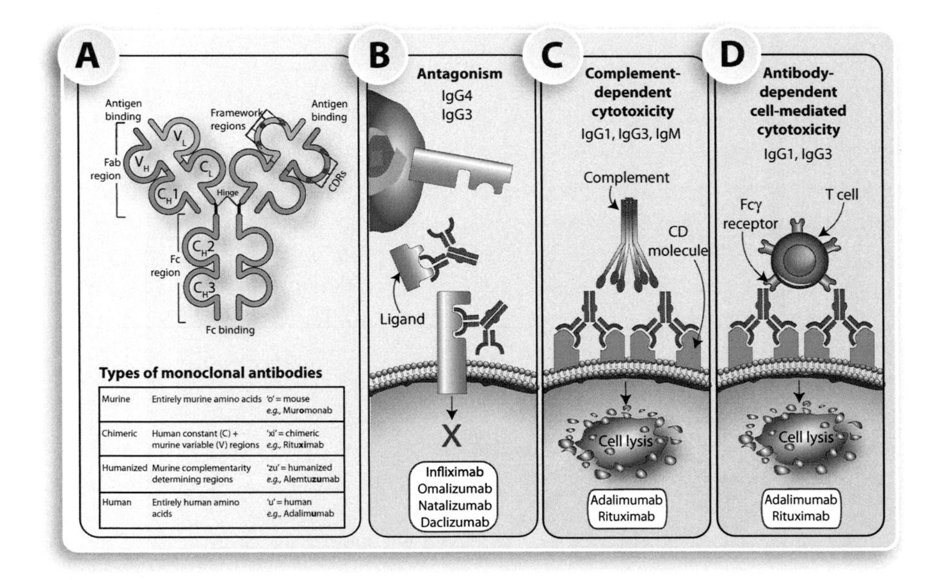

Fig. 10.4 Structure of (monoclonal) antibodies and their application in oncology. Structure of an antibody of type IgG (immunoglobin G) (**A**). Monoclonal antibodies are produced in tissue culture using hybridoma technology, where a murine B cell producing an antibody of desired specificity obtains unlimited growth potential by fusion with a myeloma cell. The type of monoclonal antibody can be deduced from elements of its name: there are original murine antibodies, chimeric antibodies with murine variable regions grafted onto human constant regions, humanized antibodies having human scaffold and murine complementarity-determining regions (CDRs) and antibodies with complete human sequences. The functions of monoclonal antibodies include antagonism (**B**) complement-dependent cytotoxicity (**C**) and antibody-dependent cell-mediated cytotoxicity (**D**)

target. In their fight against microbes, antibodies have two major functions: either they neutralize microbes by binding to their surface, in order to block the microbe's contact with host cells, or they specifically mark the microbe for the recognition of other components of the immune system. The latter are either proteins of the complement system or cytotoxic T cells, which specifically bind the constant part of the antibodies (Fc region, Fig. 10.4A). Since most antigens have multiple epitopes, a number of different B cell clones respond to the presence of the antigen and produce a polyclonal set of antibodies. Polyclonal antibodies are often a challenge for a therapeutic application, which requires high specificity, in order to minimize side effects. Thus, **in oncology only monoclonal antibodies are used for immunotherapy**.

Approved antibodies (Table 10.2) use different mechanisms:

- direct action of the antibody via receptor antagonism, e.g., blockade of agonist activity of a oncoprotein, such as the RTK HER2 (Trastuzumab, trade name "Herceptin") (Fig. 10.4B)
- immune-mediated cell killing, such as complement-dependent cytotoxicity, e.g., via binding to the B cell-specific surface protein CD20 (Rituximab) (Fig. 10.4C)
- immune-mediated cell killing, such as antibody-dependent cellular cytotoxicity, e.g., via binding to the chemokine receptor CCR4 (C-C chemokine receptor) (Mogamulizumab) (Fig. 10.4D).

The blockage of the immune checkpoint inhibiting proteins CTLA4, PD1 or CD274 (Fig. 10.5) via the antibodies Ipilimumab, Nivolumab/Pembrolizumab and Atezolizumab/Avelumab/Durvalumab, respectively, follows the first mechanism. Thus, **blocking the security mechanism for possible overactions of the immune system is boosting the cancer surveillance potential of T cells**. For some cancer patients this antibody-based immunotherapies show impressive durable responses, but unfortunately the majority of patients do not get long-term benefits, when using these immune checkpoint inhibitors as monotherapy. Importantly, the efficiency of neoantigen presentation via MHC-I proteins, which in part is based on the polymorphisms of the encoding genes *HLA-A*, *HLA-B* and *HLA-C*, correlates with the success of immune checkpoint blockage. Interestingly, unfavorable microbiome signatures of the gut are linked to poor responses to immune checkpoint blockade suggesting that a number of extrinsic factors modulate the response to immunotherapy (Sect. 8.2). Moreover, the efficiency of immune checkpoint blockade but also of other immunotherapies depend on the amount of immune cell infiltration of cancers, which is low for brain tumors and uveal melanoma (an eye cancer). In particular, the amount of professionally antigen-presenting cells, such as dendritic cells and macrophages, in non-lymphoid tissues is critical. It is low in pancreas and brain, mid-level in filtering organs, such as the kidney and liver, and highest in skin, lung and gut that are heavily exposed to the environment.

The safety and efficacy of therapeutic monoclonal antibodies largely depends on the targeted antigen. Ideally, the latter should be expressed exclusively and abundantly on the surface of cancer cells, while secreted antigens risk to prevent antibody binding to malignant tumor cells. Moreover, when complement- and antibody-dependent cytotoxicity is chosen as mechanism of action, the antigen-antibody

Table 10.2 Therapeutic monoclonal antibodies approved for use in oncology. A selection of monoclonal antibodies, which had been approved by FDA (US Food & Drug Administration) and/or EMA (European Medicines Agency) since 1997, are listed

Name (trade name)	Antigen	Antibody type	Approved indications (selection)
Rituximab (Rituxan and MabThera)	CD20	Chimeric IgG1, κ-chain	Non-Hodgkin lymphoma, CLL
Ofatumumab (Arzerra)	CD20	Human, mouse-derived IgG1, κ-chain	CLL
Obinutuzumab (Gazyva)	CD20	Humanized, glyco-engineered IgG1, κ-chain	CLL, follicular lymphoma
Ibritumomab tiuxetan (Zevalin)	CD20	Mouse IgG1, κ-chain; 90Y-containing radioimmunoconjugate	Non-Hodgkin lymphoma
Tositumomab (Bexxar)	CD20	Mouse IgG2a, λ-chain; 131I-containing radioimmunoconjugate	Non-Hodgkin lymphoma
Inotuzumab ozogamicin (Besponsa)	CD22	Humanized IgG4, κ-chain	ALL
Atezolizumab (Tecentriq)	CD274	Human, phage-derived, aglycosylated IgG1, κ-chain	Urothelial cancer and NSCLC
Avelumab (Bavencio)	CD274	Human, phage-derived IgG1, λ-chain	Merkel cell and urothelial cancer
Durvalumab (Imfinzi)	CD274	Human, transgenic mouse-derived IgG1, κ-chain	Urothelial cancer
Brentuximab vedotin (Adcetris)	CD30	Chimeric IgG1, κ-chain	Hodgkin lymphoma, systemic anaplastic large cell lymphoma
Gemtuzumab ozogamicin (Mylotarg)	CD33	Humanized IgG4, κ-chain	AML
Daratumumab (Darzalex)	CD38	Human, transgenic mouse-derived IgG1, κ-chain	Multiple myeloma
Ipilimumab (Yervoy)	CTLA4	Human, transgenic mouse-derived IgG1, κ-chain	Melanoma
Cetuximab (Erbitux)	EGFR	Chimeric IgG1, κ-chain	Colorectal cancer, head and neck cancer
Panitumumab (Vectibix)	EGFR	Human, transgenic mouse-derived IgG2, κ-chain	Colorectal cancer

(continued)

Table 10.2 (continued)

Name (trade name)	Antigen	Antibody type	Approved indications (selection)
Necitumumab (Portrazza)	EGFR	Human, phage-derived IgG1, κ-chain	Squamous NSCLC
Nimotuzumab (TheraCIM, BIOMAb-EGFR)	EGFR	Humanized IgG1, κ-chain	Glioma, head and neck, nasopharyngeal and pancreatic cancer
Trastuzumab (Herceptin)	HER2	Humanized IgG1, κ-chain	Breast, gastric and gastroesophageal junction cancer
Pertuzumab (Perjeta)	HER2	Humanized IgG1, κ-chain	Breast cancer
Ado-trastuzumab emtansine (Kadcyla)	HER2	Humanized IgG1, κ-chain	Breast cancer
Nivolumab (Opdivo)	PD1	Human, mouse-derived IgG4, κ-chain	Melanoma, Hodgkin lymphoma, squamous cell carcinoma
Pembrolizumab (Keytruda)	PD1	Humanized IgG4, κ-chain	Melanoma, Hodgkin lymphoma, NSCLC
Olaratumab (Lartruvo)	PDGFRA	Human, transgenic mouse-derived IgG1, κ-chain	Soft tissue sarcoma
Bevacizumab (Avastin)	VEGFA	Humanized IgG1, κ-chain	Non-squamous NSCLC and colorectal cancer
Ramucirumab (Cyramza)	VEGFR2	Human, phage-derived IgG1, κ-chain	NSCLC, gastric and colorectal cancer

complex should not be internalized, while the latter is desired, if the antibody delivers a radioisotope or other toxins or downregulates a protein on the surface of the cancer cell.

10.4 Immune Cell Therapies

Red blood cell transfusions are used since some 70 years as cell therapies in the context of trauma, surgery and bone marrow failure. Later, platelet transfusions as well as hematopoietic stem cell transplantation were introduced, in order to improve the survival of patients with hematological diseases. More recently, **adoptive cell transfer**, also called immune cell therapy, has emerged for the treatment of solid and hematological cancers. Adoptive cell transfer is mostly applied for T cells in an autologous fashion, i.e., patients are getting back their own cells. In contrast, in allogenic approaches the patient receives cells from a donor, who needs to match as close as possible in the HLA profile, in order to avoid graft rejections.

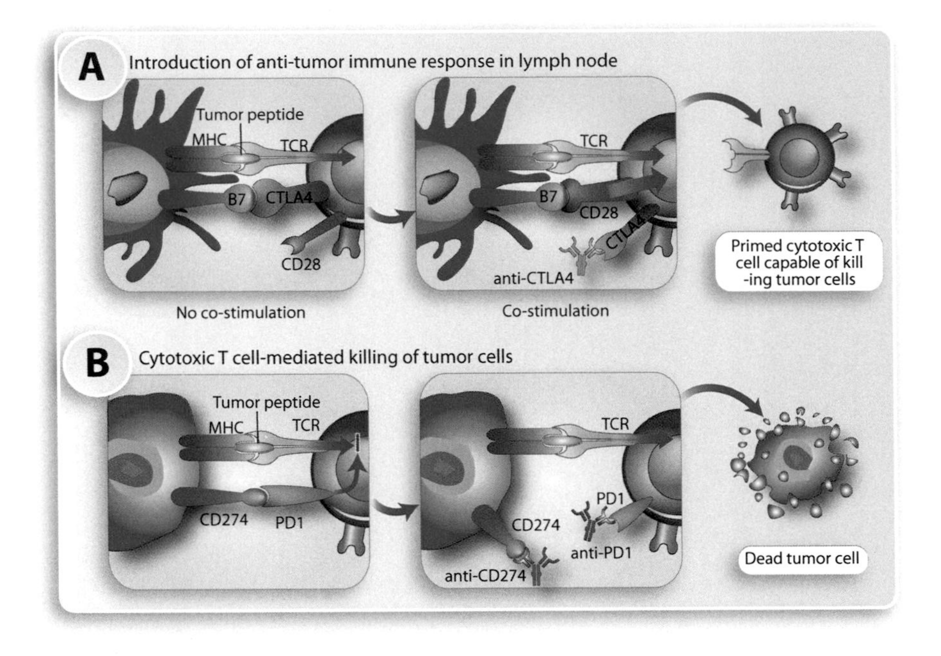

Fig. 10.5 Immune checkpoint blockade. T cell responses to malignant tumors are often inefficient due to the upregulation of inhibitory receptors of the immune checkpoint, such as CTLA4 and PD1 on the tumor-specific T cells and CD274 on the malignant tumor cells. Blocking these immune checkpoint inhibitors via anti-CTLA4 antibodies (**A**) or anti-PD1 or anti-CD274 antibodies (**B**) results in the activation of cytotoxic T cells and the induction of apoptosis to cancer cells

In a process referred to as **apheresis**, whole blood is removed from a patient or donor, selected immune cells are extracted and the remaining blood components are given back to the same person. The selected immune cells are often tumor-specific T cells, which are cultured in vitro in the presence of IL2 and other growth stimulating cytokines, in order to increase their number (Fig. 10.6A). The phase of in vitro culture allows further purification of the cells and their genetic engineering (Fig. 10.6B). When the cells are given back to the patient, they are expected to boost the natural ability of the immune system fighting against the malignant tumor.

In some solid cancers, such as melanoma, adoptive transfer of tumor-infiltrating T cells or of **T cells expressing recombinant TCRs** recognizing tumor-specific neoantigens showed impressive responses. This approach is most promising for cancers with a large mutational burden attracting a high number of TILs. In contrast, in case of other cancer types, where the number of tumor-specific T cells is rather low, this adoptive cell transfer approach is not very promising.

The therapy options of hematological malignancies, such as ALL, B cell lymphoma and multiple myeloma, had a significant breakthrough by adoptive transfer of T cells, which had been engineered with CARs (Fig. 10.7A) (Box 10.4). For example, CAR T cells targeting the B cell-specific membrane proteins CD19 and

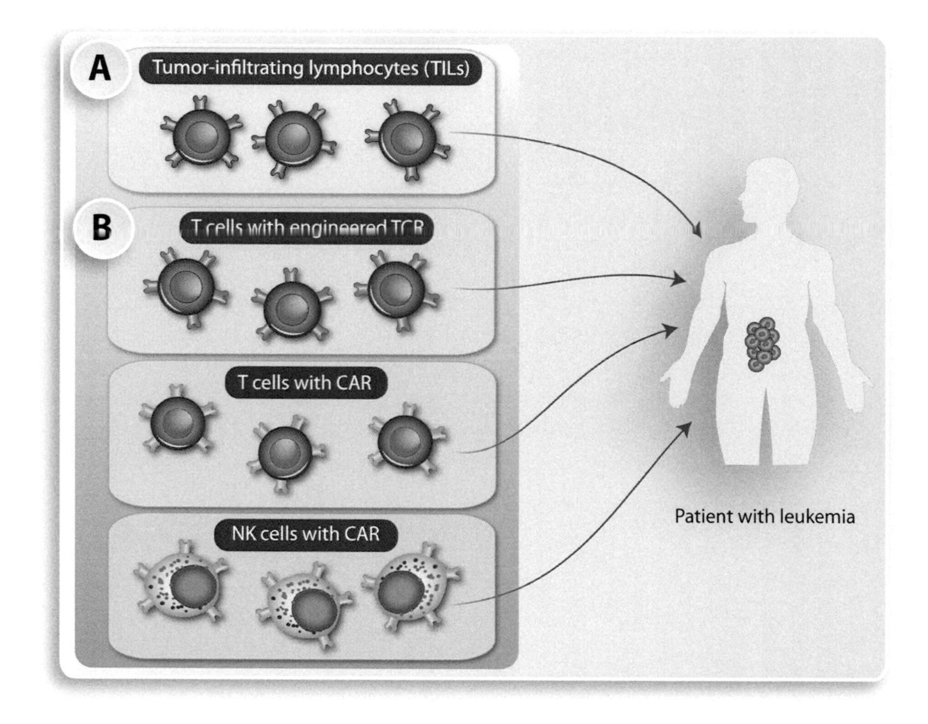

Fig. 10.6 Immune cell therapies in cancer treatment. Immune cell therapies are performed either without engineered receptors using TILs (**A**) or with T cells transduced with engineered TCR or T cells and NK cells transduced with CAR (**B**). The engineered cell types are grown in vitro before re-infusion into the patient. In all cases the cells specifically recognize malignant tumor cells and eliminate them

CD22 have been approved first in 2017 by FDA and EMA for the treatment of B cell malignancies, such as of relapsed or refractory B cell ALL, relapsed or refractory diffuse large B cell lymphoma and primary mediastinal large B cell lymphoma. Other specific target proteins on the surface of cancer cells are TNFRSF (TNF receptor superfamily member) 8 in Hodgkin lymphoma and anaplastic large cell lymphoma, TNFRSF17, SLAMF7 (SLAM family member 7) and CD38 in multiple myeloma as well as CD33, IL3RA (interleukin 3 receptor subunit alpha) and CLEC12A (C-type lectin domain family 12 member A) in AML.

Box 10.4: CAR T cells. Through the expression of CARs, T cells can be specifically directed against a given malignant tumor. CARs are recombinant proteins that are composed of two extracellular antigen-targeting Ig domains of an antibody, a so-called single-chain variable fragment (scFv), which has high specificity for a given tumor antigen, and cytoplasmic signaling domains that activate T cells, such as the ITAM (immunoreceptor tyrosine-based activation motif) domain of the TCR co-receptor CD3ς and co-stimulatory domains of the proteins CD28 and/or TNFRSF9 (Fig. 10.7B). Recombinant technologies allow to generate scFvs to any cell surface molecule, such as modified proteins, lipids, sugars and neoantigens, displaying a broad range of biochemical properties. Thus, the combination of immunotherapy and genetic engineering has created with CAR T cells a kind of "living" drugs

CAR T cells are mostly used in an autologous fashion, i.e., they derive from patient's own T cells. An individual manufacturing process is required for each patient and causes a treatment delay of at least 3 weeks. In contrast, allogeneic CAR T cells from donors are immediately available for treatment and can be produced as standardized batches at far lowers costs. Nevertheless, the latter cells bear the risk of causing life-threatening graft-versus-host disease as well as host-versus-graft

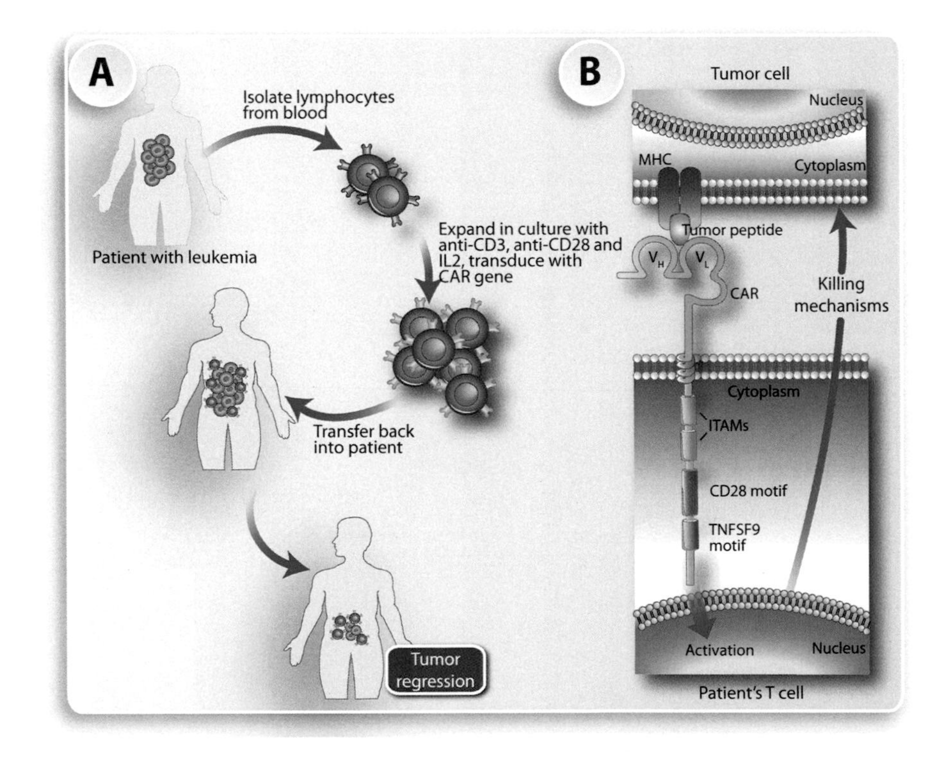

Fig. 10.7 CAR T cell therapy. T cells are isolated from the patient's blood expanded in vitro by culture in the presence of the cytokine IL2. A specific lymphodepletion is achieved by removing CD3 and CD28 positive cells via respective specific antibodies. The remaining cells are transduced with CARs and transferred back to the patient (**A**). CARs are recombinant proteins composed of two variable Ig domains specific for a tumor antigen and cytoplasmic signaling domains that activate T cells (**B**). More details are provided in the text

rejection, i.e., rapid elimination of the therapeutic cells. Therefore, HLA matching between donor and patient is of utmost importance. Furthermore, a well-known immunologic phenomenon, T cell exhaustion due to chronic antigen stimulation, can limit the efficacy of CAR T cells. Unfortunately, the present generation of CAR T cells did not show any convincing efficacy in patients with solid malignant tumors. This is related to the lack of high-level, homogeneous expression of membrane proteins on solid malignant tumors compared to normal tissue. In addition, CAR T cells are inhibited by the suppressive immune cells being part of the microenvironment of solid malignant tumors (Sect. 8.2). However, there is the chance that next-generation CAR T cells, which are engineered to overexpress the cytokine IL12 for promoting survival or enzymes for handling the tumor microenvironment, have a higher success rate.

Although NK cells belong to the innate immune system, i.e., they do not express any high-specificity antigen receptor, they are able to kill malignant tumor cells and are an integral part of our body's tumor immunosurveillance system. Different types of cancer have evolved mechanism inhibiting the function of NK cells and in this way escape from being killed. Therefore, an attractive concept is to transduce NK cells with CARs (Fig. 10.6B).

Clinical conclusion: Most malignant tumors develop due to a failure of the cancer surveillance mechanisms of our immune system. Thus, taking care on the potency of our immune system is the best way of cancer prevention. Moreover, different approaches of immunotherapy promise to be the most effective way of treating a cancer.

Further Reading

Bruni, D., Angell, H. K., & Galon, J. (2020). The immune contexture and Immunoscore in cancer prognosis and therapeutic efficacy. *Nature Reviews Cancer, 20,* 662–680.

Carter, P. J., & Lazar, G. A. (2018). Next generation antibody drugs: pursuit of the 'high-hanging fruit'. *Nature Reviews Drug Discovery, 17,* 197–223.

Demaria, O., Cornen, S., Daeron, M., Morel, Y., Medzhitov, R., & Vivier, E. (2019). Harnessing innate immunity in cancer therapy. *Nature, 574,* 45–56.

Depil, S., Duchateau, P., Grupp, S. A., Mufti, G., & Poirot, L. (2020). 'Off-the-shelf' allogeneic CAR T cells: development and challenges. *Nature Reviews Drug Discovery, 19,* 185–199.

Ferreira, L. M. R., Muller, Y. D., Bluestone, J. A., & Tang, Q. (2019). Next-generation regulatory T cell therapy. *Nature Reviews Drug Discovery, 18,* 749–769.

Garner, H., & de Visser, K. E. (2020). Immune crosstalk in cancer progression and metastatic spread: a complex conversation. *Nature Reviews Immunology, 20,* 483–497.

Havel, J. J., Chowell, D., & Chan, T. A. (2019). The evolving landscape of biomarkers for checkpoint inhibitor immunotherapy. *Nature Reviews Cancer, 19,* 133–150.

Keenan, T. E., Burke, K. P., & Van Allen, E. M. (2019). Genomic correlates of response to immune checkpoint blockade. *Nature Medicine, 25,* 389–402.

Larson, R. C., & Maus, M. V. (2021). Recent advances and discoveries in the mechanisms and functions of CAR T cells. *Nature Reviews Cancer, 21,* 145–161.

Majzner, R. G., & Mackall, C. L. (2019). Clinical lessons learned from the first leg of the CAR T cell journey. *Nature Medicine, 25,* 1341–1355.

Smith, C. C., Selitsky, S. R., Chai, S., Armistead, P. M., Vincent, B. G., & Serody, J. S. (2019). Alternative tumor-specific antigens. *Nature Reviews Cancer, 19,* 465–478.

Weber, E. W., Maus, M. V., & Mackall, C. L. (2020). The emerging landscape of immune cell therapies. *Cell, 181,* 46–62.

Chapter 11
Architecture of Cancer Therapies

Abstract The classical therapy of solid cancers is surgery, which is often combined with radiation therapy and cytotoxic chemotherapy. The latter is an economic way of cancer treatment but often causes side effects. Nowadays, the application of next-generation methods, such as exome and whole genome sequencing, enables a detailed analysis of the molecular complexity of cancer in general as well as a focus on the individual case. In particular, understanding the role of kinases as drivers in the tumorigenesis process allowed the design of inhibitory molecules, many of which are successfully applied as anti-cancer drugs with high precision. The combination of genomic analysis with the increasing number of specific anti-cancer drugs and the expanding variety of immune therapies permits a more precise treatment of each individual cancer case.

Keywords Surgery · Radiation therapy · Cytotoxic chemotherapy · Small-molecule inhibitors · Targeted therapy · Precision oncology

11.1 Classical Cancer Treatments

At present, approximately 5% of the population in Western countries have the diagnosis of some type of cancer, i.e., they are obtaining treatment or they are in fear of a possible live-threatening relapse. The overall survival (Box 11.1) critically depends on the type of cancer and whether only a primary malignant tumor is detected or already many metastatic tumors are present. Furthermore, a number of other factors contribute to cancer survival (Sect. 11.3).

Box 11.1: Definition of cancer cure. The diagnosis of cancer is a challenge to the personal integrity of the patient. The typical first question of the patient is of the chance to survive the disease. The cure of cancer is defined by years of survival with three timepoints serving as cornerstones:

C. Carlberg and E. Velleuer, *Cancer Biology: How Science Works*,
https://doi.org/10.1007/978-3-030-75699-4_11

- **two-year survival**: the chances to die due to cancer are the highest within the first two years after cancer diagnosis either due to progression, relapse or to side effects of the treatment
- **five-year survival**: patients who survive the first five years after diagnosis of most adult cancers are defined as cured
- **ten-year survival**: ten years after the initial diagnosis children with cancer are discharged from aftercare, i.e., they are defined as cured.

Beside overall survival, disease-free survival is understood as the time between cancer diagnosis and the time without a relapse. In contrast, the event-free survival is the time interval between initial diagnosis and the new onset of cancer related symptoms. Relative survival is statistically adjusted to the normal life expectancy and controlled to age, gender and other influencing factors.

The classical therapy of solid cancers begins with the local control of the cancer. In most cases, this is surgery, which is traditionally performed by using thin knives or by lasers, when higher precision is required. In open surgery there is one large cut to remove the malignant tumor, some healthy tissue and the nearby lymph nodes (at least the sentinel lymph node). In contrast, in minimally invasive surgery a laparoscope for monitoring the progress of the surgery is inserted by a small cut while via other small cuts the malignant tumor and surrounding healthy tissue are removed. The recovery time from the latter is obviously shorter. Sometimes, surgery is needed to remove parts of the tumor to enable normal body function (debulking), e.g., if the malignant tumor compresses important vessels or other anatomical structures. Surgery performed with laser or with the support of robots can be quite accurate but are limited to only few cancer types.

Some malignant tumors are localized in regions of the body, such as the brain, which are difficult or impossible to reach by surgery without severely harming healthy surrounding tissues. In these cases, often radiotherapy as local control is applied, where high energy X-rays or protons are used to kill cancer cells via the induction of unrepairable DNA damage. Dead cancer cells are then removed by macrophages using efferocytosis. However, the dying of cancer cells can take weeks so that the full effect of radiotherapy is the earliest seen 6 weeks after. Radiotherapy is either applied by an external beam or via a device that is placed inside the patient's body. The radiation source is either solid within a capsule and placed close to the malignant tumor, such as head and neck, breast, cervix, prostate and eye (brachytherapy) or the treatment is systemically, e.g., through monoclonal antibodies coupled with a radioisotope (Table 10.2) or a radioisotope with well-established tropism, such as I_{131} for the treatment of thyroid cancers or I_{131}-metaiodobenzylguanidine for the treatment of neuroblastoma (Fig. 2.3). However, radiotherapy bears the risk of transforming healthy cells, i.e., it may create new malignant tumors.

Most types of solid cancers become life-threatening when they have formed metastases (Chap. 9). **When the latter have already been detected by clinical imaging methods, there is often no realistic chance of cure**. Since metastases can spread long time before the diagnosis of a primary malignant tumor, most patients receive in addition to surgical removal of their primary malignant tumor (i.e., **local control**) systemic therapy, such as chemotherapy, in order to target possible undetected metastases (i.e., **systemic control**) (Box 11.2). These treatments are called "adjuvant" when they are performed after surgery or radiotherapy and "neoadjuvant" when they are given before cancer removal. In the latter case the aim is to shrink the cancer before surgery and to measure the individual's response rate to chemotherapy. Neoadjuvant concepts are mostly applied in protocols for the treatment of pediatric cancers. Since chemotherapy is interfering with wound healing, it can be started only few weeks after surgery. Thus, **with rapidly growing cancers chemotherapy is often done first**.

Box 11.2: Categories of cancer therapies and cancer treatment architecture. Cancer treatment is subdivided into local and systemic control:

- **local control** is conducted via surgery, radiotherapy, hyperthermia or photodynamic therapy
- depending on the type of cancer **systemic control** is achieved with classical intravenous treatment, such as chemotherapy, immunotherapy, hormone therapy, anti-metabolite therapy and precision therapy.

Choosing the right drug at the right time during systemic control is of critical importance. During **induction phase**, the goal is to kill as many malignant cells as possible. In this phase, patients suffer from many treatment side effects, since the doses of the drugs are high (Fig. 11.1). During **consolidation phase,** less toxic drugs are applied to keep cancer cell number low. Finally, in some cancers, **maintenance therapy** is applied, in order to kill cancer stem cells who survived induction therapy. These drugs are mostly orally admitted and do not cause direct side effects and are given for a long time. Resistance of the cancer against chemotherapy can be reduced by admitting multiple drugs at the same time.

At present, cancer treatment architectures are multi-modal:

- protocols stratify patients based on their cancer stage (Box 4.2)
- optimization protocols randomize between standard and alternative treatment, where either the efficacy of new drugs (superior) or the reduction of therapy related toxicity (non-inferior) is tested

- aftercare protocols support the patient with possible side effects of the treatment and aim to catch a possible relapse or secondary cancer as early as possible

The principle of conventional cytotoxic chemotherapy is to kill fast-growing cancer cells by drugs, such as anthracylins, cisplatin and anti-metabolites that primarily interfere with DNA replication. Chemotherapy is given by infusion at a maximally tolerated dose (Fig. 11.1), i.e., **small changes in drug concentrations can make the difference between substantial cancer regression and intolerable side effects**. However, patients show interindividual differences in drug metabolism, which can be monitored on the level of the SNPs within the genes encoding for the respective enzymes. Thus, drug dosing adapted to pharmaco-kinetic

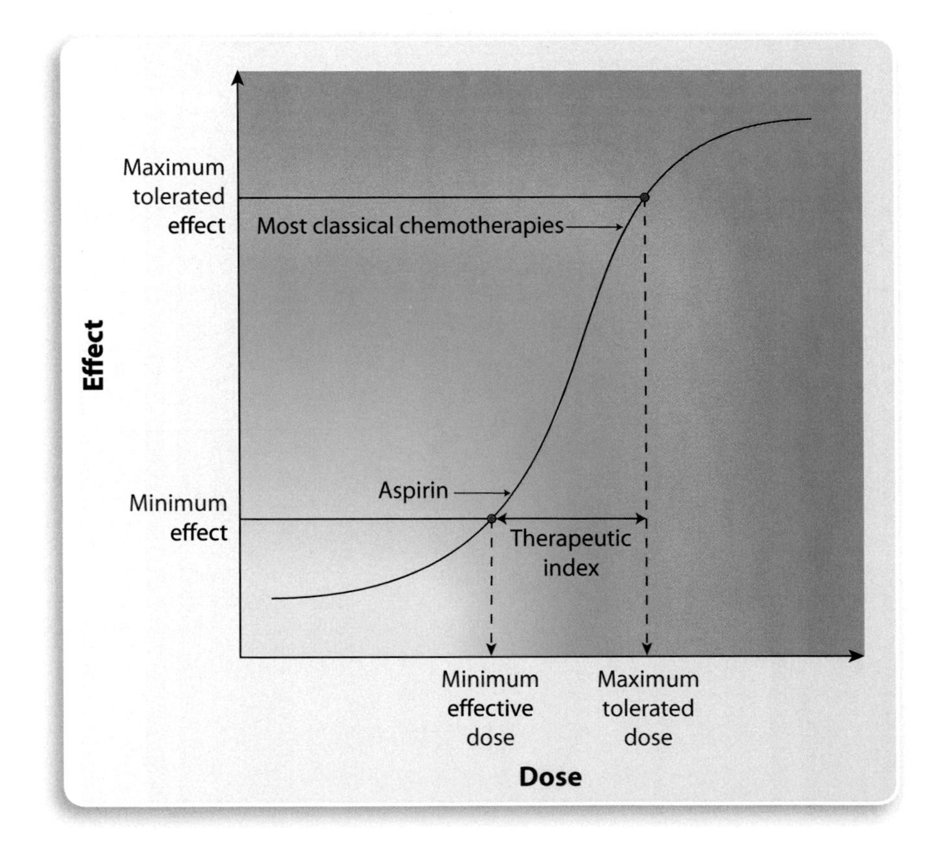

Fig. 11.1 Conventional cytotoxic chemotherapy. The dose response curve illustrates that cytotoxic chemotherapeutic drugs are given at maximally tolerated doses, in order to get largest effect on cancer cell killing

and genetic parameters has the chance to obtain optimal results. Nevertheless, **cytotoxic chemotherapeutic drugs often act in a rather unspecific way and affect also healthy growing cells**, so that, e.g., hairs are lost. Therefore, cytotoxic chemotherapy causes a number of side effects, such as nephrotoxicity (kidney damage), neurotoxicity (nerve damage), nausea and vomiting, ototoxicity (hearing loss), electrolyte disturbance as well as transient aplasia of the bone marrow.

The overall survival benefits of many types of cytotoxic chemotherapics are limited. Although they often initially eliminate the dominant drug-sensitive cell clone of a heterogeneous cancer and significantly reduce the malignant tumor burden for a limited time, they allow the accelerated growth of drug-resistant, more aggressive cancer cell clones in a competition-free rich environment. Thus, **the selective pressure of chemotherapy allows drug-resistant cancer cells to prosper**. This results in an even more rapid disease progression compared to tumor growth before treatment and therefore does not necessarily prolong the time of survival (Fig. 11.2). Therefore, it is of utmost importance that therapies address the completeness of all cancer cell clones. Mostly, this can be achieved by combining different drugs in different treatment intervals (sequential therapy). For example, in many childhood

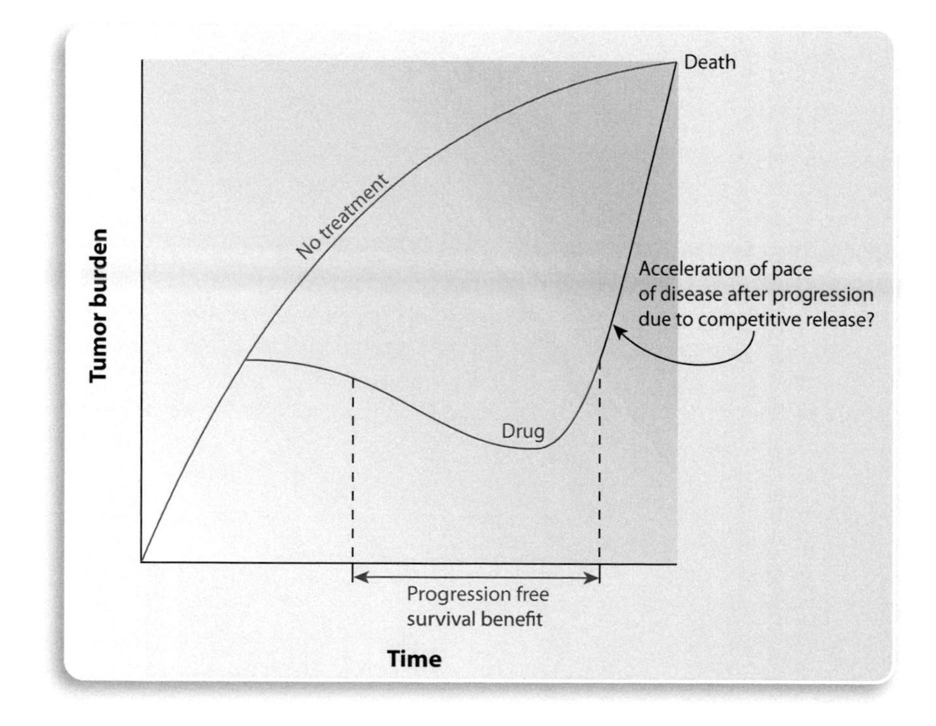

Fig. 11.2 Treatment of resistant cancer cell subclones. Treated and untreated patients often have similar overall survival times, if the treatment does not eliminate aggressive subclones that are able to accelerate their growth in a competition-free tumor environment

cancers, during the induction phase cytotoxic chemotherapy is applied aiming on a rapid reduction of the tumor burden. Next, during consolidation phase, different drugs or the same drugs like in induction phase but at lower doses are applied to stabilize cancer size by stopping exponential growth, while in a maintenance phase preferentially orally admitted drugs are applied. The aim of maintenance therapy is to catch the cancer stem cells in their metastatic niches when they wake up from dormancy. Thus, **maintenance therapy can prevent macrometastasis or relapse** (Sect. 9.3). For example, in breast cancer hormone therapy with anti-estrogens or with monoclonal antibodies like Trastuzumab (Sect. 10.3) are maintenance therapies and can be applied for years.

11.2 Targeted Therapies

Standard cytotoxic chemotherapy is a rather unspecific therapy by killing both malignant and healthy rapidly growing cells via the inhibition of DNA replication (Sect. 11.1). In contrast, **targeted cancer therapies have a specific molecular target**. For example, the treatment of $HER2^+$ breast cancers with a monoclonal antibody, such as Trastuzumab, against the RTK HER2 (ERBB2) is very specific and targeted (Sect. 10.3). In addition, a cancer vaccine that is designed against a neoantigen on the surface of a cancer cells is targeting cytotoxic T cells very specifically to the respective malignant tumor (Sect. 10.2). Also, the use of CARs engineered to recognize specific surface molecules and directing cytotoxic T cells toward them is highly specific and targeted (Sect. 10.4). However, the original meaning of targeted cancer therapy is the treatment with a small synthetic molecule that easily passes molecular membranes and binds specifically to its target protein. **Often, these molecules are inhibitors that bind to proteins having a pro-cancer action**. Hormone therapy is another example for targeted therapy, since the applied drug, such as the anti-estrogen tamoxifen, is binding specifically to a molecular target like the nuclear receptor ESR1. In this way, tamoxifen blocks the binding of the natural ESR1 ligand estrogen, which would stimulate the proliferation of breast cancer. Inhibitors of the enzyme CYP19A1 (cytochrome P450 family 19 subfamily A member 1, also called aromatase), which blocks the conversion of androgens into estrogen, have a similar effect like anti-estrogens.

Most targeted therapies address one or several of the 10 hallmarks of cancer (Sect. 2.4). Based on the hallmark concept, the dysregulation of one or several signaling pathways can explain the tumorigenesis process. A perfect toolbox for targeted cancer therapy would contain for each hallmark a number of specific approved drugs (Fig. 11.3). When the main driver of a given cancer has been identified and a specific inhibitory drug is available, targeted therapy can be performed. However, often this driver hallmark is based on more than one dysregulated pathway, i.e., the application of a single drug will result only in a partial reduction of the hallmark activities. In this case the clinical responses will most likely only be transitory

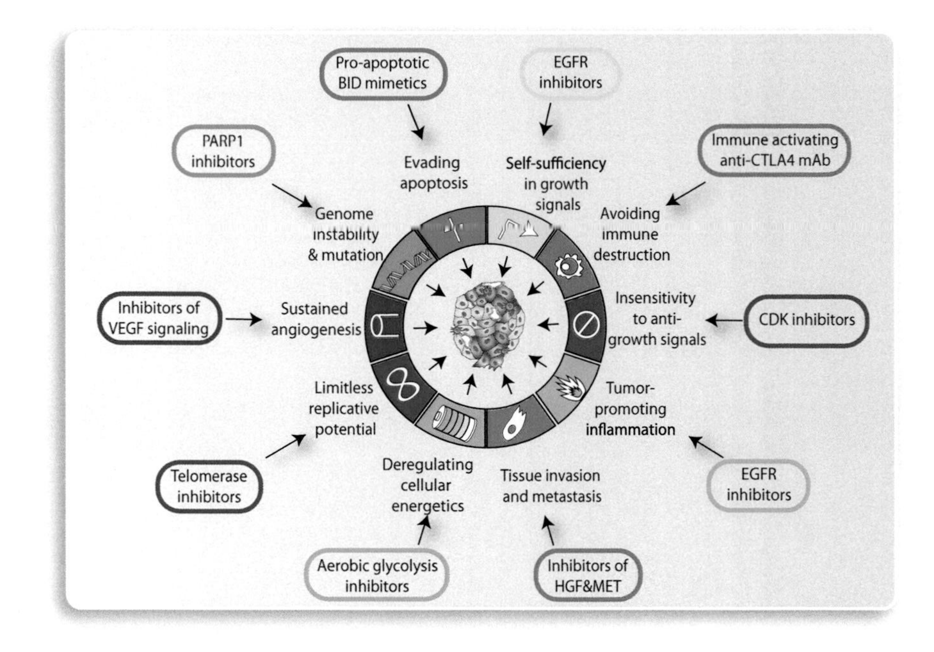

Fig. 11.3 Drug targeting the hallmarks of cancer. For most of the 10 hallmarks of cancer drugs have been developed that interfere with the respective acquired capabilities. BID = BH3 interacting domain death agonist, mAB = monoclonal antibody, MET = MET proto-oncogene, receptor tyrosine kinase

followed by a relapse (Fig. 11.2). Therefore, **a combination therapy using two or more different drugs is essential**.

Protein kinases regulate key cellular processes, such as proliferation, survival and migration. When they are dysregulated, they contribute to the various hallmarks of cancer. Accordingly, the most common protein domain encoded by cancer driver genes is the protein kinase domain being encoded by more than 100 of the 576 known genes (Sect. 5.3). This represents more than 20% of some 500 protein kinases encoded by the human genome. Druggable cancer drivers belong to the following protein families:

- nearly half of the 58 known members of the RTK family of transmembrane proteins
- five members of the TGFβ superfamily
- eight non-receptor tyrosine kinases
- many kinases related to the signaling cascades of RAS-RAF1-MAPK and PI3K-mTOR
- many kinases that are related to cell cycle control and DNA damage signaling, such as CDK4, ATM and ATR.

Cancer genome projects (Sect. 5.3) indicated that genes encoding for a number of well-characterized protein kinases are mutated at high frequencies in one or several cancers, such as *BRAF* in thyroid cancer (60%), skin cutaneous melanoma (51%), multiple myeloma (15%) and colorectal cancer (10%) and *EGFR* in glioblastoma (27%) and lung adenocarcinoma (12%). Thus, some kinases are drivers in multiple types of malignant tumors, while others, such as KIT (KIT proto-oncogene, receptor tyrosine kinase) in AML and PDGFRA in glioblastoma, are unique to specific cancers. Furthermore, key regulatory pathways, such as 76% of the RTK-RAS-RAF1 cascade and 25% of PI3K-mTOR signaling, contain mutations in protein kinase genes. Most of the driver kinases represent oncogenes, i.e., they are activated by mutations. Thus, **the most efficient therapy seems to be the use of small-molecule inhibitors**.

Since all protein kinases are enzymes that use ATP as a co-substrate, they all possess an ATP binding domain via which they are druggable. Therapeutic molecules are available already for some 25% of all cancer driver kinases, but not all are yet approved. For example, the small-molecule inhibitor imatinib (brand name "Gleevec") targets the BCR-ABL1 fusion protein (Sect. 2.3) in Philadelphia chromosome-positive CML and ALL as well as gain-of-function mutations of the oncogenes *KIT* and *PDGFRA* in gastrointestinal stromal tumors, chronic eosinophilic leukemia, systemic mastocytosis and the myelodysplastic syndrome. Similarly, the BRAF inhibitor vemurafenib had been developed for BRAF-V600E$^+$ thyroid carcinoma, the EGFR inhibitors gefitinib, erlotinib, afatinib, brigatinib and icotinib for lung cancer and cetuximab for colorectal cancer and the ALK (anaplastic lymphoma kinase) inhibitor crizotinib to treat cancers with *ALK* gene translocations. However, since only a small proportion of the cancers carry the respective mutations, **a response to targeted therapy can only be expected when the mutation is confirmed**.

Transcription factors, such as those encoded by the key cancer genes, *MYC*, *JUN*, *FOS* and *TP53*, are challenging to target, since they do not have any enzymatic activity that may be inhibited (for oncoproteins) or activated (for tumor suppressor proteins) by small molecules. Moreover, although there is already a reasonable number of drugs that inhibit overactive oncogenes, it is difficult to design molecules that re-activate inactivated tumor suppressor genes or replace their lost function. One approach is to use inhibitors of oncogenes that are natural antagonists to tumor suppressor actions, such as the inhibitor olaparib of the oncogene product PARP1 (poly(ADP) ribose polymerase 1) in case of ovarian cancers that are deficient of the tumor suppressor genes *BRCA1* or *BRCA2*. Since in respective cancer cells the HR DNA repair pathway (Sect. 4.3) does not work, PARP1 inhibition results in their death. Thus, **non-mutated genes and proteins are targeted by inhibitory drugs, in order to selectively eliminate transformed cells**. Cancer genome projects have identified high-frequency alterations in DNA damage response genes, which provides drugs that target DNA repair pathways, such as PARP1 inhibitors, wider applications. Using the same principles, mutations disabling the tumor suppressor gene *PTEN* can be balanced by inhibitors of the kinase AKT, *CDKN2A* mutations by CDK4 inhibitors and *APC* mutations by inhibitors of CTNNB1.

An alternative approach is to affect the regulation of tumor suppressor genes via modulators of chromatin modifiers (Sect. 6.1). For example, HDAC inhibitors reactivate the transcription of tumor suppressor genes, such as *CDKN1A*, by increasing histone acetylation, but they also mediate the acetylation of non-histone proteins, such as p53, and stabilize their activity. In this way, they have a wide impact on cancer cells and can induce apoptosis, cell cycle arrest and many other anti-cancer actions. Currently, five HDAC inhibitors vorinostat, romidepsin, belinostat, panobinosta and valproic acid are approved for treatment of different types cancer such as for leukemia, T cell lymphomas and multiple myeloma. Valproic acid, which primarily used for the treatment of epilepsy, is thereby redefined for the treatment of some childhood cancers. As many other HDAC inhibitors are currently under testing in clinical trials, it is expected that more HDAC inhibitors will be developed for the treatment of solid tumors, in particular for brain cancers.

The contribution of altered epigenetic states to the phenotype of cancer cells suggests that **epigenetic cancer therapies** have a clinical impact. For example, genes encoding for epigenetic modifiers, such as KMTs and KDMs (Fig. 11.4), are frequent drivers in a larger range of cancer types. Moreover, in contrast to genetic mutations,

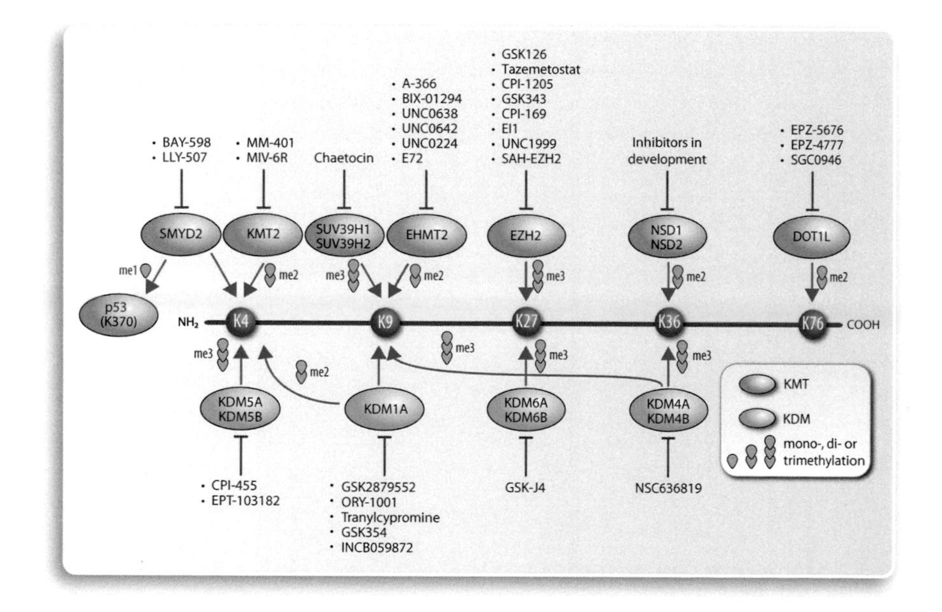

Fig. 11.4 Targeting KMT and KDM mutations. Inhibitors of KMTs (red) and KDMs (blue) are indicated that affect histone 3 lysines K4, K9, K27, K36, and K79 (blue). KMT inhibitors bind either within the SAM pocket, within the substrate pocket or at allosteric sites of the KMT protein. Respective inhibitors of DOT1L and EZH2 are already in clinical trials. Chaetocin is a non-selective inhibitor of KMTs, such as SUV39H1 and SUV39H2. KDM1A is a FAD-dependent KDM, which can be inhibited by molecules blocking its co-factor binding site. The KDM1A inhibitors tranylcypromine, GSK2879552, INCB059872, and ORY-1001 are in clinical trials. Most KDMs carry a catalytic domain, which can be inhibited by iron-chelating molecules

epigenetic modifications are largely reversible. Basically all molecules designed for epigenetic cancer therapy are inhibitors of enzymes affected by gain-of-function mutations. Since histone methylation marks have a far more selective function than histone acetylation marks, **KMT and KDM inhibitors promise to be more specific and may be less toxic than HDAC inhibitors or even DNMT inhibitors**. Several KMT inhibitors have been developed, and the H3K27-KMT EZH2 inhibitor tazemetostat has already been approved for the treatment of epithelioid sarcoma and follicular lymphoma. Since EZH2 is the catalytic core of the PRC2 complex that also recruits DNMTs, EZH2 inhibitors may link both epigenetic repression mechanisms. Moreover, the inhibition of EZH2 results in reduced levels of H3K27me3 marks, upregulation of silenced genes and inhibition of the growth of cancer cells with EZH2 gain-of-function mutations or overexpression. KDMs use FAD (flavin adenine dinucleotide), α-ketoglutarate or Fe(II) as co-factors and offer in this way a number of options for their inhibition. However, the catalytic domain of most KDMs is structurally highly conserved, which is a challenge for the design of specific KDM inhibitors.

DNMT inhibitors are either nucleoside analogs that after incorporation into the DNA covalently trap DNMTs or non-nucleoside analogs that directly bind to the catalytic region of DNMTs. These molecules prevent DNA methylation leading to reduced promoter hypermethylation and re-expression of silenced tumor suppressor genes. The two nucleoside analogs azacitidine and decitabine have already been approved more than 30 years ago. At present, **epigenetic drugs are used primarily for the treatment of hematologic cancers** in combination with cytotoxic chemotherapies, targeted therapies and immune checkpoint inhibitors. This broadens the response rates among patients and provides a potential for the treatment of solid tumors.

Finally, modulators of chromatin modifiers are not only effective in the therapy of different forms of cancer, but also in the prevention of malignant tumor formation. Interestingly, a number of natural, food-derived compounds, such as epicatechin from green tea, resveratrol from grapes or curcumin from curcuma, are known to modulate the activity of chromatin modifiers. Thus, **ingredients of healthy diet have the potential to prevent cancer by keeping the activity of chromatin modifiers under control**.

11.3 Precision Oncology

The diversity of cancer types in combination with the understanding that the disease is based on an individual set of driver mutations made oncology to a pioneer of **precision medicine** (Box 11.3). The kind of "one size fits all" approaches of cytotoxic chemotherapy (Sect. 11.1) leading to severe side effects are getting over time replaced by (or are at least combined with) more precise therapies using drugs and treatments based on a molecular diagnosis of the individual's cancer case. These are on one hand approaches to disrupt pathways being essential for cancer growth and survival (pathway-based targeted therapies, Sect. 11.2) and on the other hand attempts to push the patient's immune system for a better response against malignant tumor cells (immunotherapies, Chap. 10) (Fig. 11.5). Targeted therapies build on the understanding of driver genes and proteins forming pathways, which are dysregulated and/or dysfunctional in a given individual cancer case. Accordingly, large-scale data, such as whole genome sequence information or gene expression profiles, are essential for a precise diagnosis. This allows choosing the most appropriate drug, in case

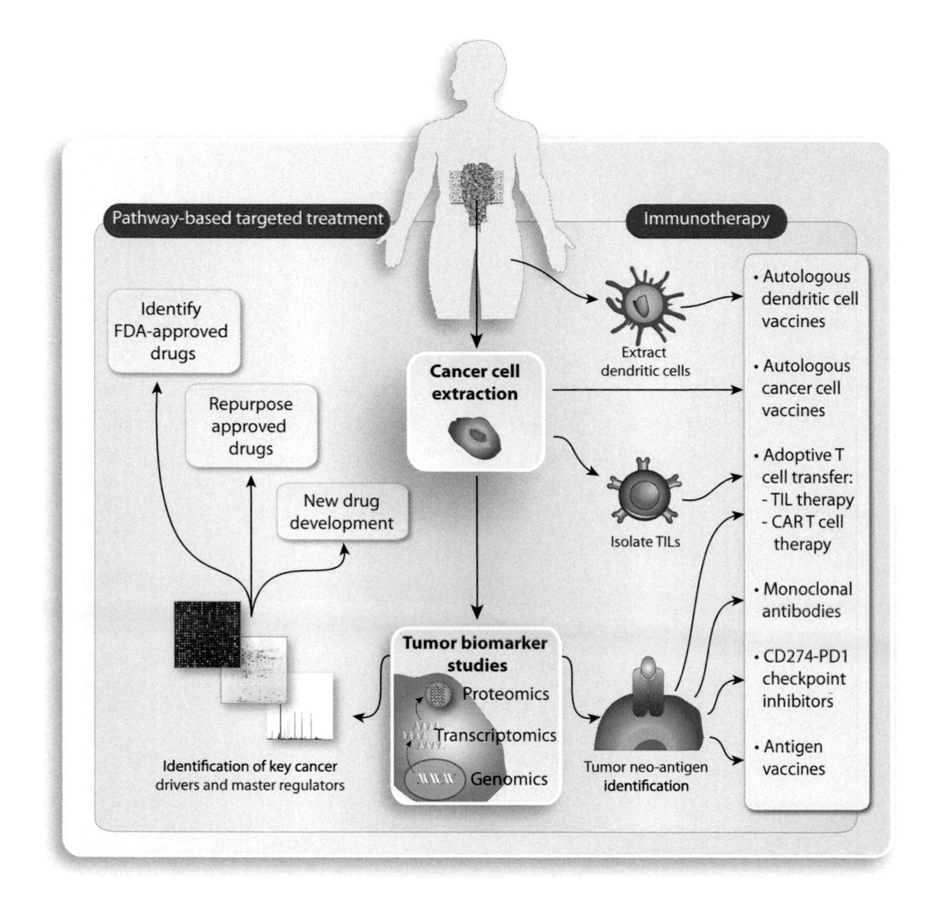

Fig. 11.5 Precision therapy of cancer. There are two major approaches for precision oncology: pathway-based targeted therapies (**left**) and immunotherapies (**right**). In both cases, cancer cell samples are analyzed by a number of methods, such as next-generation sequencing, gene expression profiling and proteomics. For pathway-based targeted treatments the main focus is the detection of key drivers and master regulators of the cancer, in order to identify possible approved drugs. Immunotherapy approaches include cell-based therapies using patient-derived dendritic cells and T cells, tumor neoantigen-based vaccines and biologics, such as monoclonal antibodies and CAR T cells. Tumor antigen information is often based on next-generation sequencing data and allows most precise decisions for the appropriate therapy protocol

an approved molecule is already available. This concept is followed in the German experimental protocol "INFORM" (www.dkfz.de/en/inform/index.html) focusing on childhood cancer with remarkable results. Furthermore, for the decision, which of the in Chap. 10 presented precise immunotherapies may be applied, large-scale data as well as the molecular understanding of cancer immunology are crucial.

Box 11.3: Precision medicine. The aim of precision medicine is to design the treatment of a disease based on the patient's own genetic background and other molecular and cellular data, i.e., the therapy is tailored to the characteristics of the patient. This is in clear contrast to a "one drug fits all" approach and acknowledges interindividual difference. However, the idea is not new but nowadays supported by clearly improved diagnostic technologies. Often precision medicine is also referred as personalized medicine, but the latter implies the misleading assumption that for each patient a unique drug or medical device is applied. In contrast, precision medicine rather classifies individuals into subgroups that differ in their susceptibility to a given disease and/or their response to a particular treatment. Accordingly, preventive or therapeutic interventions can then be concentrated on those who will benefit from it. Precision medicine is often based on large-scale data on the genome, epigenome, transcriptome, proteome and/or metabolome of the patient. The success of the precise therapy is tracked as closely as possible, often also using large-scale data or molecular biomarkers. Precision medicine provides health care providers with more insight on the patient's environment, lifestyle and heredity with its effects on health and disease. **This shifts the emphasis in medicine from reaction to prevention, allows a prediction on the individual's susceptibility to disease, improves disease detection, customizes disease-prevention strategies and leads to the prescription of more effective drugs with less side effects**. Ideally, this should reduce costs and will significantly improve patient care

Big biology cancer genome projects, such as TCGA (Sect. 5.3), provided a detailed genomic characterization of malignant tumors suggesting a new taxonomy of cancer types complements current histology-based classifications (Sect. 1.2, Box 4.2). This suggests that molecular features rather than the tissue type will determine future therapies, i.e., patients whose malignant tumors carry the same mutation, e.g., *BRAF* V600, are eligible for treatment with BRAF inhibitor vemurafenib regardless of its cancer type. Nevertheless, **for most patients a precision oncology approach is not yet part of routine care**. However, the genetic profiling for *HER2* gene amplification in breast cancer and for mutations in the genes *KIT*, *CEBPA*, *FLT3* and *NPM1* (nucleophosmin 1) in AML belongs to the standard. Although genetic alterations of certain oncogenes, such as *MYC* and *RAS*, occur in 28 and 16% of all cancers and that of the tumor suppressor gene *TP53* even in 50% of all malignant tumors, they are difficult to target directly and lack approved drugs.

Another aspect of precision oncology is the detection of germline mutations. For example, in 12% of prostate cancer patients mutations in the tumor suppressor genes *BRCA1*, *BRCA2*, *PALB2*, *ATM* and *CHEK2* are found, which all can be treated by PARP1 inhibitors. Similarly, defects in the MMR DNA repair system based on germline, somatic and epigenetic mutations are targetable with immune checkpoint inhibitor blockade and are detected by cancer/healthy tissue matched sequencing. In general, **genome-driven precision oncology is able to learn from each individual patient**, when respective data are shared in open access databases (Box 11.4). This will allow the treating physicians to interpret the individual genomic alterations of new patients and to identify best treatment options.

Box 11.4: Data sharing in oncology. The value of high-quality interpretations of individual genomic variants is becoming increasingly clear. Accordingly, the community of medical genetics has created a number of databases, such as ClinVar (www.ncbi.nlm.nih.gov/clinvar) and GCG (www.gdc.cancer.gov), and clinical knowledge bases, such as OncoKB (www.oncokb.org), MyCancerGenome (www.mycancergenome.org), Cancer Genome Interpreter (www.cancergenomeinterpreter.org), CANDL (www.candl.osu.edu), CIViC (www.civic.genome.wustl.edu) and the Personalized Cancer Medicine Knowledge Base (www.pct.mdanderson.org)

Taken together, cancer therapy is changing due to the detailed molecular description of individual cancers and the availability of potent and selective small-molecule inhibitors as well as monoclonal antibodies. However, the knowledge how to use the powerful tool of precision oncology needs to be implemented in the routine care of cancer patients. **The diagnosis of cancer destroys the personal integrity and is a huge burden for the patient**, his/her family and the team of physicians involved in the treatment. To achieve cure, an individual understanding of the biology of tumorigenesis and the mechanisms of treatment is indispensable. Moreover, many other factors contribute to the effectiveness of the treatment (Fig. 11.6).

Clinical conclusion: Cancer biology teaches us how precious life is and how well our body is coordinated. As all of us have only one life to live, we should take responsibility. One patient pointed out: "Even though I'm diagnosed with a cancer predisposition syndrome and fought already some cancers, I take each day as the chance to create something new and to move on."

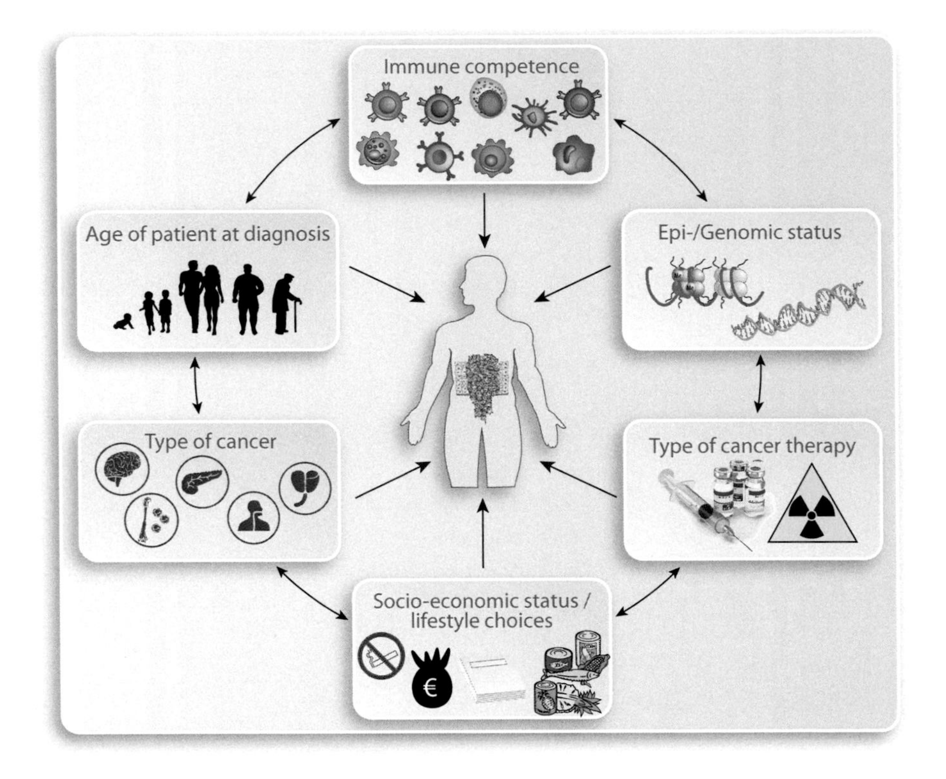

Fig. 11.6 Factors contributing to cancer survival. Survival of cancer depends on many factors that interact with each other. Young patients are diagnosed with different types of cancer than adults and tolerate treatment much better (Box 6.4). Moreover, patients with cancer predisposition syndromes fight against multiple types of cancer. The immune competence of the individual is critical for cancer control and therapy. The type of cancer therapy is very important for survival, since cytotoxic chemotherapeutics many acute side effects and contribute to a substantial number of long-term organ damages. Interindividual genomic variations contribute positively or negatively to therapy tolerance. Finally, the socio-economic status and lifestyle choices of the patient impacts overall health and therapy commitment

Further Reading

Clarke, M. F. (2019). Clinical and therapeutic implications of cancer stem cells. *The New England Journal of Medicine, 380,* 2237–2245.

Dugger, S. A., Platt, A., & Goldstein, D. B. (2018). Drug development in the era of precision medicine. *Nature Reviews Drug Discovery, 17,* 183–196.

Grandori, C., & Kemp, C. J. (2018). Personalized cancer models for target discovery and precision medicine. *Trends Cancer, 4,* 634–642.

Hyman, D. M., Taylor, B. S., & Baselga, J. (2017). Implementing genome-driven oncology. *Cell, 168,* 584–599.

Kuenzi, B. M., & Ideker, T. (2020). A census of pathway maps in cancer systems biology. *Nature Reviews Cancer, 20,* 233–246.

Sweet-Cordero, E. A., & Biegel, J. A. (2019). The genomic landscape of pediatric cancers: Implications for diagnosis and treatment. *Science, 363,* 1170–1175.

Vasan, N., Baselga, J., & Hyman, D. M. (2019). A view on drug resistance in cancer. *Nature, 575,* 299–309.

Glossary

Adaptive therapy the application of cancer treatment in a manner that quickly responds to changes in the disease rather than following a fixed protocol with the goal of managing the cancer and maintaining a limited tumor burden, rather than attempting to totally eliminate the disease

Adenoma a benign tumor composed of epithelial cells

Adjuvant therapy is therapy that is given in addition to the primary or initial therapy to maximize its effectiveness such as chemotherapy after surgery

Amplification a genetic alteration producing a large number of copies of a small segment (less than a few Mb) of the genome

Angiogenesis the process of forming vascular conduits including veins, arteries and lymphatics

Benign tumor an abnormal proliferation of cells driven by at least one mutation in an oncogene or tumor suppressor gene. These cells are not invasive (i.e. they cannot penetrate the basement membrane lining them), which distinguishes them from malignant cells

Cachexia a complex syndrome associated causing ongoing muscle loss that is not entirely reversed with nutritional supplementation

Cancer stem cells are cancer cells that possess characteristics associated with normal stem cells such as the ability to give rise to all cell types found in a particular cancer sample

Carcinoma a type of malignant tumor composed of epithelial cells

Chemokines a family of small cytokine-like proteins that induce directed chemotaxis to responsive neighboring cells i.e., they act as chemotactic cytokines

Chemotherapy is a type of cancer treatment that uses one or more anti-cancer drugs

Chromatin the molecular substance of chromosomes being a complex of genomic DNA and histone proteins

Chronic inflammation long-term inflammation lasting for prolonged periods of several months to years

C. Carlberg and E. Velleuer, *Cancer Biology: How Science Works*, https://doi.org/10.1007/978-3-030-75699-4

Clonal mutation a mutation that exists in the vast majority of the neoplastic cells within a malignant tumor

Cytokines a category of small proteins (~5–20 kDa) that are important in cell signaling involved in autocrine paracrine and endocrine signaling as immunomodulating agents

Damage-associated molecular patterns (DAMPs) also known as alarmins are molecules often released by stressed cells undergoing necrosis that act as endogenous danger signals to promote and exacerbate inflammatory responses

DNA methylation the covalent addition of a methyl group to the C5 position of cytosine

Driver gene mutation (driver) a mutation that directly or indirectly confers a selective growth advantage to the cell in which it occurs

Driver gene a gene that contains driver gene mutations or is expressed aberrantly in a fashion that confers a selective growth advantage

Ectoderm the outermost layer of the three embryonic germ layers that gives rise to the epidermis like skin, hair and eyes, as well as the nervous system

Embryogenesis also called embryonic development i.e., the process by which the embryo forms and develops. In mammals, the term is use exclusively to the early stages of prenatal development, whereas the terms fetus and fetal development describe later stages

Endoderm the innermost layer of the three embryonic germ layers that gives rise to the epithelia of the digestive and respiratory systems such as liver, pancreas and lungs

Epigenetics the study of heritable changes in gene function that do not involve changes in the DNA sequence. Epigenetic mechanisms include the covalent modifications of DNA and histones

Epigenome the complete set of epigenetic modifications across an individual's genome

Epimutation heritable change in the chromatin state at a given position or region. In the context of cytosine methylation epimutations are defined as changes in the methylation status of a single cytosine or of a region or cluster of cytosines. Epimutations do not necessarily imply changes in gene expression

Epitope specific portion of the antigen specifically recognized by a TCR or BCR

Exome The collection of exons in the human genome. Exome sequencing generally refers to the collection of exons that encode proteins

Gatekeeper a gene that when mutated, initiates tumorigenesis. Examples include *RB1*, mutations of which initiate retinoblastomas, and *VHL*, whose mutations initiate renal cell carcinomas

Gene expression process by which information from a gene is used in the synthesis of a functional gene product. These products are often proteins but can also be ncRNAs

Genome the complete haploid DNA sequence of an organism comprising all coding genes and far larger non-coding regions. The genome of all 400 tissues and cell types of an individual is identical and constant over time (with the exception of cancer cells)

Genome-wide association study (GWAS) a study that aims to identify genetic loci (mostly SNPs) associated with an observable trait disease or condition

Genotype complete heritable genetic identity

Germline variants variations in sequences observed in different individuals. Two randomly chosen individuals differ by ~20,000 genetic variations distributed throughout the exome and some 4 million SNPs in total

Haploinsufficiency occurs when a single copy of the wildtype allele in heterozygous combination with a variant allele is insufficient to produce the wildtype phenotype

Healthspan the duration of disease-free physiological health within the lifespan of an individual. In humans this corresponds to the period of high cognitive abilities, immune competence and peak physical condition

HLA typing process for identifying the HLA receptor allele of a particular tissue

Homozygous deletion deletion of both copies of a genomic region

Human leukocyte antigen (HLA) a protein encoded by genes that determine an individual's capacity to respond to specific antigens or reject transplants from other individuals. *HLA* genes encode for MHC proteins

Immune checkpoint blockage immune checkpoints are regulators of the immune system being crucial for self-tolerance. Immune checkpoint blocking proteins such as CTLA4 and PD1, are targets for cancer immunotherapy

Immunosenescence is the age-associated gradual deterioration of the immune system in particular of the adaptive immune system. This involves capacity of the host to respond to infections, to develop of long-term immune memory and to perform efficient immune surveillance

Insertion or deletion (indel) a mutation due to small insertion or deletion of one or a few nucleotides

Karyotype display of the chromosomes of a cell on a microscopic slide used to evaluate changes in chromosome number as well as structural alterations of chromosomes

Kinase a protein that catalyzes the addition of phosphate groups to other molecules such as proteins or lipids. These proteins are essential to nearly all signal transduction pathways

Liquid tumors tumors composed of hematopoietic (blood) cells such as leukemias. Though lymphomas generally form solid masses in lymph nodes, they are often classified as liquid tumors because of their derivation from hematopoietic cells and their ability to travel through lymphatics

Malignant tumor an abnormal proliferation of cells driven by mutations in oncogenes or tumor suppressor genes that has already invaded their surrounding stroma. It is impossible to distinguish an isolated benign tumor cell from an isolated malignant tumor cell. This distinction can be made only through examination of tissue architecture

Massive parallel sequencing high-throughput approach to DNA sequencing using the concept of massively parallel processing. It is also called next-generation sequencing or deep sequencing

Mesoderm the middle layer of the three embryonic germ layers that gives rise to the muscle cartilage, bone, blood, connective tissue etc

Metastatic tumor a malignant tumor that has migrated away from its primary site such as to draining lymph nodes or another organ

Missense mutation A single-nucleotide substitution (e.g. C to T) that results in an amino acid substitution (e.g., histidine to arginine)

Mitogen a signaling molecule such as a peptide or small protein, that stimulates a signal transduction cascade inducing progression of the cell cycle and mitosis

Mitosis cell division that results in two daughter cells each having the same number and kind of chromosomes as the parent nucleus

Non-communicable disease a disease that is not transmissible directly from one person to another such as autoimmune diseases, CDVs, most cancers, diabetes and Alzheimer's disease

Non-homologous end joining a major double-strand break repair pathway that does not rely on sequence homology and can result in small insertions and deletions at the site of repair

Nonsense mutation a single-nucleotide substitution (e.g. C to T) that results in the production of a stop codon

Non-synonymous mutation a mutation that alters the encoded amino acid sequence of a protein. These include missense nonsense, splice site, translation start, translation stop, and indel mutations

Neoantigens antigens specific to cancer cells

Oncogene a gene that when activated by mutation, increases the selective growth advantage of the cell in which it resides

Passenger mutation (passenger) a mutation that has no direct or indirect effect on the selective growth advantage of the cell in which it occurred

Pathogen-associated molecular patterns (PAMPs) small molecular motifs derived from microbes such a lipopolysaccharides. They are recognized by toll-like receptors and other pattern recognition receptors on the surface of cells of the innate immune system

Phenotype the set of observable characteristics of an individual resulting from the interaction of its genotype with the environment

Plasticity the reversibility of epigenetic marks on DNA and proteins

Primary malignant tumor the original malignant tumor at the site where cancer growth was initiated. This can be defined for solid malignant tumors but not for liquid tumors

Rearrangement a mutation that juxtaposes nucleotides that are normally separated such as those on two different chromosomes

RNA sequencing (RNA-seq) a method using massive parallel sequencing to reveal the presence and quantity of RNA in a biological sample at a given moment

Senescence also called biological aging the gradual deterioration of functional characteristics. It can refer either to cellular senescence or to senescence of the whole organism

Signal transduction cascade the process by which a chemical or physical signal is transmitted through a cell membrane as a series of molecular events such as protein phosphorylation catalyzed by protein kinases. Mostly, signal transduction cascades end in the activation of a transcription factor or a chromatin modifier

Single base substitution A single-nucleotide substitution (e.g. C to T) relative to a reference sequence or, in the case of somatic mutations, relative to the germline genome of the person with a cancer

Single nucleotide polymorphism (SNP) a substitution of a single nucleotide at a specific position in the genome which is present to some appreciable degree within a population (e.g., more than 1%)

Solid malignant tumors malignant tumors that form discrete masses such as carcinomas or sarcomas

Somatic mutations mutations that occur in any non-germ cell of the body after conception such as those that initiate tumorigenesis

Stem cells can differentiate into other cell types and also divide in self-renewal to produce more of the same type of stem cells. There are embryonic stem cells which are isolated from the inner cell mass of blastocysts, and adult stem cells, which are found in various tissues

Stroma the connective functionally supportive framework of a biological cell, tissue or organ

Trait a distinguishing quality or characteristic belonging to a person

Transcription factors proteins that sequence-specifically bind to genomic DNA. Our genome encodes approximately 1600 transcription factors referred to as *trans*-acting factors, since they are not encoded by the same genomic regions, which they are controlling. Accordingly, the process of transcriptional regulation by transcription factors is often called *trans*-activation

Transcriptome the complete set of all transcribed RNA molecules of a tissue or cell type. It significantly differs between tissues and depends on extra and intracellular signals

Translocation a specific type of rearrangement where regions from two nonhomologous chromosomes are joined

Translocation breakpoints locations where two fragments of chromosome(s) are joined subsequent to chromosomal translocation

Tumor antigens any antigen produced by the malignant tumor cell typically in the setting of enriched or specific expression relative to normal tissue(s)

Tumor suppressor gene a gene that when inactivated by mutation, increases the selective growth advantage of the cell in which it resides

Printed in the United States
by Baker & Taylor Publisher Services